# THE WORLD BOOK ENCYCLOPEDIA OF
# PEOPLE AND PLACES

# 2
# C-F

WORLD
BOOK

a Scott Fetzer company
Chicago
www.worldbookonline.com

For information about other World Book publications,
visit our website at http://www.worldbookonline.com
or call 1-800-WORLDBK (1-800-967-5325).

For information about sales to schools and libraries, call
1-800-975-3250 (United States);
1-800-837-5365 (Canada).

**Library of Congress Cataloging-in-Publication Data**

The World Book encyclopedia of people and places.
    v. cm.
    Summary: "A 7-volume illustrated, alphabetically arranged
set that presents profiles of individual nations and other
political/geographical units, including an overview of history,
geography, economy, people, culture, and government of each.
Includes a history of the settlement of each world region
based on archaeological findings; a cumulative index; and Web
resources"--Provided by publisher.
    Includes index.
    ISBN 978-0-7166-3758-5
    1. Encyclopedias and dictionaries. 2. Geography--
Encyclopedias. I. World Book, Inc. Title: Encyclopedia of
people and places.
    AE5.W563 2011
    030--dc22
                                                2010011919
This edition ISBN: 978-0-7166-3760-8

Printed in Hong Kong by Toppan Printing Co. (H.K.) LTD
3rd printing, revised, August 2012

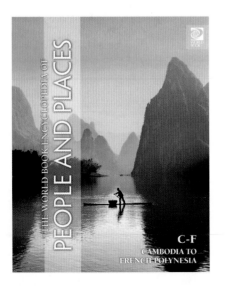

**Cover image:**
Li River, Guilin, China

© SIME/eStock Photo

# CONTENTS

The world has 196 independent countries and about 50 dependencies. An independent country controls its own affairs. Dependencies are controlled in some way by independent countries. In most cases, an independent country is responsible for the dependency's foreign relations and defense, and some of the dependency's local affairs. However, many dependencies have complete control of their local affairs.

By 2010, the world's population was nearly 7 billion. Almost all of the world's people live in independent countries. Only about 13 million people live in dependencies.

Some regions of the world, including Antarctica and certain desert areas, have no permanent population. The most densely populated regions of the world are in Europe and in southern and eastern Asia. The world's largest country in terms of population is China, which has more than 1.3 billion people. The independent country with the smallest population is Vatican City, with only about 830 people. Vatican City, covering only 1/6 square mile (0.4 square kilometer), is also the smallest in terms of size. The world's largest nation in terms of area is Russia, which covers 6,601,669 square miles (17,098,242 square kilometers).

Every nation depends on other nations in some way. The interdependence of the entire world and its peoples is called *globalism*. Nations trade with one another to earn money and to obtain manufactured goods or the natural resources that they lack. Nations with similar interests and political beliefs may pledge to support one another in case of war. Developed countries provide developing nations with financial aid and technical assistance. Such aid strengthens trade as well as defense ties.

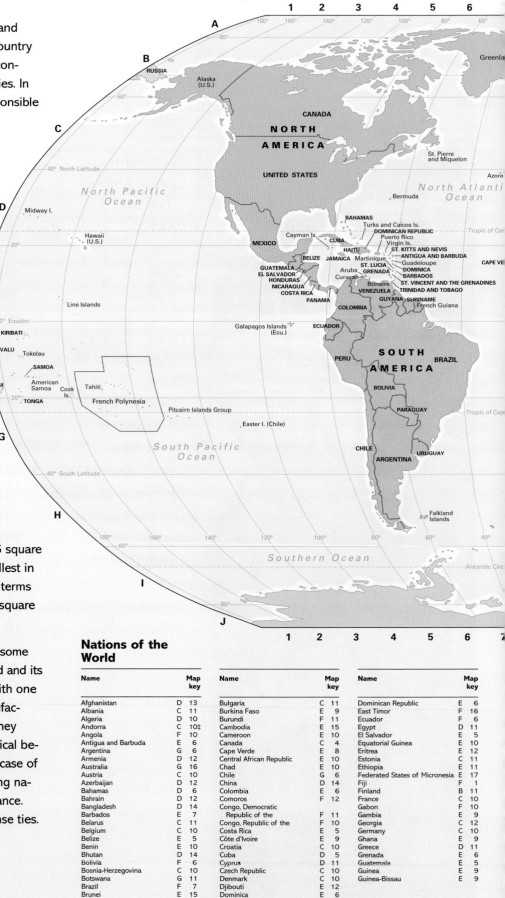

## Nations of the World

| Name | Map key | Name | Map key | Name | Map key |
|---|---|---|---|---|---|
| Afghanistan | D 13 | Bulgaria | C 11 | Dominican Republic | E 6 |
| Albania | C 11 | Burkina Faso | E 9 | East Timor | F 16 |
| Algeria | D 10 | Burundi | F 11 | Ecuador | F 6 |
| Andorra | C 10‡ | Cambodia | E 15 | Egypt | D 11 |
| Angola | F 10 | Cameroon | E 10 | El Salvador | E 5 |
| Antigua and Barbuda | E 6 | Canada | C 4 | Equatorial Guinea | E 10 |
| Argentina | G 6 | Cape Verde | E 8 | Eritrea | E 12 |
| Armenia | D 12 | Central African Republic | E 10 | Estonia | C 11 |
| Australia | G 16 | Chad | E 10 | Ethiopia | E 11 |
| Austria | C 10 | Chile | G 6 | Federated States of Micronesia | E 17 |
| Azerbaijan | D 12 | China | D 14 | Fiji | F 1 |
| Bahamas | D 6 | Colombia | E 6 | Finland | B 11 |
| Bahrain | D 12 | Comoros | F 12 | France | C 10 |
| Bangladesh | D 14 | Congo, Democratic | | Gabon | F 10 |
| Barbados | E 7 | Republic of the | F 11 | Gambia | E 9 |
| Belarus | C 11 | Congo, Republic of the | F 10 | Georgia | C 12 |
| Belgium | C 10 | Costa Rica | E 5 | Germany | C 10 |
| Belize | E 5 | Côte d'Ivoire | E 9 | Ghana | E 9 |
| Benin | E 10 | Croatia | C 10 | Greece | D 11 |
| Bhutan | D 14 | Cuba | D 5 | Grenada | E 6 |
| Bolivia | F 6 | Cyprus | D 11 | Guatemala | E 5 |
| Bosnia-Herzegovina | C 10 | Czech Republic | C 10 | Guinea | E 9 |
| Botswana | G 11 | Denmark | C 10 | Guinea-Bissau | E 9 |
| Brazil | F 7 | Djibouti | E 12 | | |
| Brunei | E 15 | Dominica | E 6 | | |

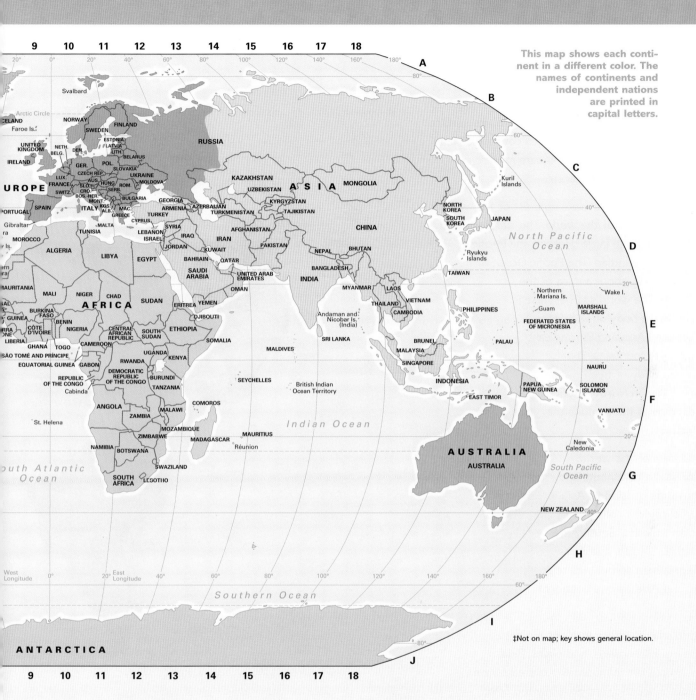

This map shows each continent in a different color. The names of continents and independent nations are printed in capital letters.

‡Not on map; key shows general location.

| Name | Map key | | Name | Map key | | Name | Map key | | Name | Map key | | Name | Map key | |
|---|---|---|---|---|---|---|---|---|---|---|---|---|---|---|
| Guyana | E | 7 | Lebanon | D | 11 | Namibia | G | 10 | St. Vincent and the Grenadines | E | 6 | Taiwan | D | 16 |
| Haiti | E | 6 | Lesotho | G | 11 | Nauru | F | 18 | Samoa | F | 1 | Tajikistan | D | 14 |
| Honduras | E | 5 | Liberia | E | 9 | Nepal | D | 14 | San Marino | C | 10‡ | Tanzania | F | 11 |
| Hungary | C | 10 | Libya | D | 10 | Netherlands | C | 10 | São Tomé and Príncipe | E | 10 | Thailand | E | 15 |
| Iceland | B | 9 | Liechtenstein | C | 10‡ | New Zealand | G | 18 | Saudi Arabia | D | 12 | Togo | E | 9 |
| India | D | 13 | Lithuania | C | 11 | Nicaragua | E | 5 | Senegal | E | 9 | Tonga | F | 1 |
| Indonesia | F | 16 | Luxembourg | C | 10 | Niger | E | 10 | Serbia | C | 10 | Trinidad and Tobago | E | 6 |
| Iran | D | 12 | Macedonia | C | 11 | Nigeria | E | 10 | Seychelles | F | 12 | Tunisia | D | 10 |
| Iraq | D | 12 | Madagascar | F | 12 | Norway | B | 10 | Sierra Leone | E | 9 | Turkey | D | 11 |
| Ireland | C | 9 | Malawi | F | 11 | Oman | E | 12 | Singapore | E | 15 | Turkmenistan | D | 13 |
| Israel | D | 11 | Malaysia | E | 15 | Pakistan | D | 13 | Slovakia | C | 11 | Tuvalu | F | 1 |
| Italy | C | 10 | Maldives | E | 14 | Palau | E | 16 | Slovenia | C | 11 | Uganda | E | 11 |
| Jamaica | E | 6 | Mali | D | 9 | Panama | E | 5 | Solomon Islands | F | 18 | Ukraine | C | 11 |
| Japan | D | 16 | Malta | D | 10 | Papua New Guinea | F | 17 | Somalia | E | 12 | United Arab Emirates | D | 12 |
| Jordan | D | 11 | Marshall Islands | E | 18 | Paraguay | G | 7 | South Africa | G | 11 | United Kingdom | C | 9 |
| Kazakhstan | C | 13 | Mauritania | D | 9 | Peru | F | 6 | Spain | C | 9 | United States | C | 4 |
| Kenya | E | 11 | Mauritius | G | 12 | Philippines | E | 16 | Sri Lanka | E | 14 | Uruguay | G | 7 |
| Kiribati | F | 1 | Mexico | D | 4 | Poland | C | 10 | Sudan | E | 11 | Uzbekistan | D | 14 |
| Korea, North | C | 16 | Moldova | C | 11 | Portugal | D | 9 | Sudan, South | E | 11 | Vanuatu | F | 18 |
| Korea, South | D | 16 | Monaco | C | 10‡ | Qatar | D | 12 | Suriname | E | 7 | Vatican City | C | 10‡ |
| Kosovo | C | 11 | Mongolia | C | 15 | Romania | C | 11 | Swaziland | G | 11 | Venezuela | E | 6 |
| Kuwait | D | 12 | Montenegro | C | 10 | Russia | C | 13 | Sweden | B | 10 | Vietnam | E | 15 |
| Kyrgyzstan | C | 13 | Morocco | D | 9 | Rwanda | F | 11 | Switzerland | C | 10 | Yemen | E | 12 |
| Laos | E | 15 | Mozambique | F | 11 | St. Kitts and Nevis | E | 6 | Syria | D | 11 | Zambia | F | 11 |
| Latvia | C | 11 | Myanmar | D | 14 | St. Lucia | E | 6 | | | | Zimbabwe | G | 11 |

# PHYSICAL WORLD MAP

The surface area of the world totals about 196,900,000 square miles (510,000,000 square kilometers). Water covers about 139,700,000 square miles (362,000,000 square kilometers), or 71 percent of the world's surface. Only 29 percent of the world's surface consists of land, which covers about 57,200,000 square miles (148,000,000 square kilometers).

Oceans, lakes, and rivers make up most of the water that covers the surface of the world. The water surface consists chiefly of three large oceans—the Pacific, the Atlantic, and the Indian. The Pacific Ocean is the largest, covering about a third of the world's surface. The world's largest lake is the Caspian Sea, a body of salt water that lies between Asia and Europe east of the Caucasus Mountains. The world's largest body of fresh water is the Great Lakes in North America. The longest river in the world is the Nile in Africa.

The land area of the world consists of seven continents and many thousands of islands. Asia is the largest continent, followed by Africa, North America, South America, Antarctica, Europe, and Australia. Geographers sometimes refer to Europe and Asia as one continent called Eurasia.

The world's land surface includes mountains, plateaus, hills, valleys, and plains. Relatively few people live in mountainous areas or on high plateaus since they are generally too cold, rugged, or dry for comfortable living or for crop farming. The majority of the world's people live on plains or in hilly regions. Most plains and hilly regions have excellent soil and an abundant water supply. They are good regions for farming, manufacturing, and trade. Many areas unsuitable for farming have other valuable resources. Mountainous regions, for example, have plentiful minerals, and some desert areas, especially in the Middle East, have large deposits of petroleum.

6

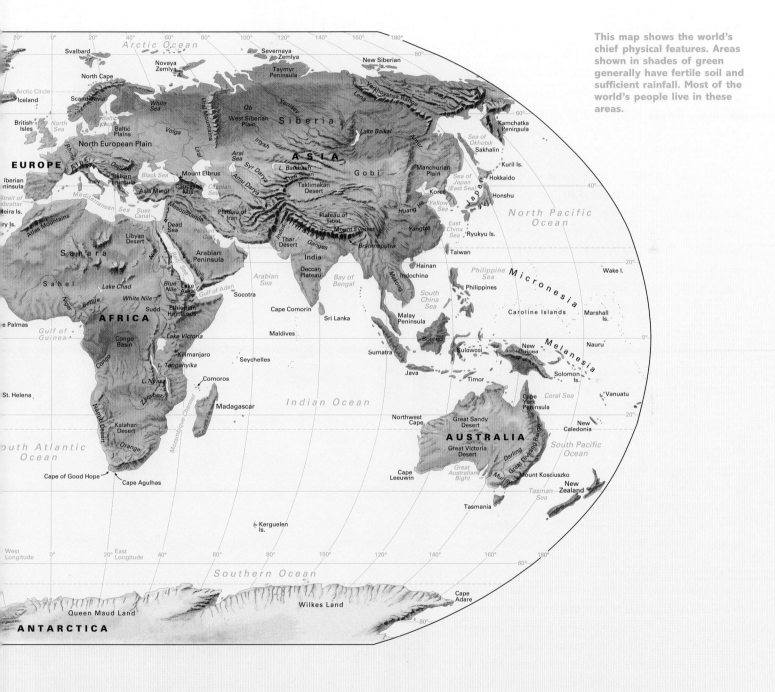

Arctic Ocean

20°  0°  20°  40°  60°  80°  100°  120°  140°  160°  180°

Svalbard
Severnaya Zemlya
New Siberian Is.
North Cape
Novaya Zemlya
Taymyr Peninsula
80°
Arctic Circle
Iceland
Scandinavia
White Sea
Ob
West Siberian Plain
S i b e r i a
Yenisey
Lena
Yerkhoyansk Range
60°
Kamchatka Peninsula
British Isles
North Sea
Baltic Plains
Ural Mountains
Volga
Sea of Okhotsk
Sakhalin
North European Plain
Baltic Sea
Irtysh
Kuril Is.
Rhine
Ural
A S I A
Aral Sea
L. Balkhash
Gobi
Manchurian Plain
Hokkaido
Sea of Japan (East Sea)
EUROPE
Alps
Danube
Black Sea
Mount Elbrus
Syr Darya
Tian Shan
Honshu
Iberian Peninsula
Italy
Balkan Peninsula
Caucasus Mts.
Amu Darya
Taklimakan Desert
Korea
Yellow Sea
40°
Asia Minor
Caspian Sea
Manchurian Plain
Strait of Gibraltar
Mediterranean Sea
Suez Canal
Mesopotamia
Plateau of Iran
Plateau of Tibet
Huang He
East China Sea
Ryukyu Is.
North Pacific Ocean
Madeira Is.
Dead Sea
Persian Gulf
Indus
Himalaya
Mount Everest
Yangtze
Canary Is.
Atlas Mountains
Red Sea
Arabian Peninsula
Thar Desert
Ganges
Brahmaputra
Taiwan
20°
S a h a r a
Libyan Desert
Nile
India
Deccan Plateau
Bay of Bengal
Mekong
Hainan
Philippine Sea
Wake I.
Sahel
Lake Chad
Blue Nile
Lake Assal
Gulf of Aden
Arabian Sea
Indochina
Micronesia
Niger
Benue
White Nile
Socotra
Cape Comorin
Philippines
South China Sea
Cape Palmas
AFRICA
Sudd
Ethiopian Highlands
Sri Lanka
Malay Peninsula
Caroline Islands
Marshall Is.
0°
Gulf of Guinea
Congo Basin
Lake Victoria
Maldives
Borneo
Nauru
Congo
Kilimanjaro
Seychelles
Sumatra
Sulawesi
New Guinea
Melanesia
St. Helena
L. Tanganyika
L. Nyasa
Comoros
Java
Timor
Solomon Is.
Vanuatu
Zambezi
Madagascar
Indian Ocean
Coral Sea
20°
South Atlantic Ocean
Namib Desert
Mozambique Channel
Northwest Cape
Great Sandy Desert
Cape York Peninsula
New Caledonia
Kalahari Desert
Orange
AUSTRALIA
Great Victoria Desert
Darling
Great Dividing Range
Mount Kosciuszko
South Pacific Ocean
Cape of Good Hope
Cape Agulhas
Cape Leeuwin
Great Australian Bight
Murray
New Zealand
Tasman Sea
Tasmania

West Longitude  0°  20° East Longitude  40°  60°  80°  100°  120°  140°  160°  180°
60°
Kerguelen Is.

Southern Ocean
80°
Cape Adare
Queen Maud Land
Wilkes Land
ANTARCTICA

This map shows the world's chief physical features. Areas shown in shades of green generally have fertile soil and sufficient rainfall. Most of the world's people live in these areas.

Cambodia, known during the 1970's and 1980's as Kampuchea, is a Southeast Asian country that borders Thailand, Laos, and Vietnam. The Mekong River flows through the eastern part of the country, creating fertile valleys ideally suited to the production of rice. The land along the Mekong is a patchwork of rice fields. Most Cambodians live on the river's fertile plains or near the *Tonle Sap* (Great Lake) and Tonle Sap River northwest of Phnom Penh, the nation's capital.

The majority of the Cambodian people are Khmer, one of the oldest groups in Southeast Asia. They speak the Khmer language, which has its own alphabet. A large Chinese community also lives in Cambodia. Most of the people living in Cambodia are Buddhists.

Cambodia's economy has traditionally been based on agriculture, particularly the production of rice and corn. In the past, the country also produced large quantities of rubber. However, many farms and rubber plantations were destroyed during the Vietnam War (1957–1975) and Cambodia's civil wars of the 1970's and 1980's. As a result, the nation's agricultural production has decreased sharply.

Since the mid-1990's, a large percentage of Cambodia's income has come from garment manufacturing. Tourism also has grown in importance.

## Early history

About 100 A.D., people in the southern part of what is now Cambodia established the kingdom of Funan, one of the greatest early powers of Southeast Asia. Funan gradually declined, and by the 600's a Khmer kingdom called Chenla had arisen north of Funan. Chenla broke up in the 700's.

Between the 800's and 1400's, the Khmer ruled a powerful Hindu-Buddhist kingdom in Cambodia. In Angkor, the capital of this empire, the Khmer built hundreds of beautiful stone temples, hospitals, and palaces. They also constructed roads, reservoirs, and canals. The Khmer empire prospered, and by the 1100's it commanded much of what is now Laos, Vietnam, and Thailand.

Internal and external conflicts eventually weakened the empire, however. In 1431, Thai forces cap-

# CAMBODIA

tured Angkor, and the Khmer abandoned the city. But an independent Khmer kingdom, with its capital near what is now Phnom Penh, survived until the mid-1800's. In 1863, the French, who occupied southern Vietnam, made Cambodia a protectorate.

## Modern history

In 1953, France granted Cambodia its independence, and two years later Cambodia's leader, King Norodom Sihanouk, gave up the throne to enter politics. He took the title of prince and became prime minister in 1955 and head of state in 1960.

In 1970, Lieutenant General Lon Nol, a member of the reigning government, overthrew Sihanouk. Lon Nol declared Cambodia a republic, and in 1971 he assumed full control of the government. Lon Nol's hold, however, was weakened by several factors. The United States, charging that North Vietnam had installed troops and supplies in Cambodia for use in the Vietnam War, began a series of bombing raids on Cambodia. Meanwhile, a Cambodian Communist organization called the *Khmer Rouge* had been growing stronger. In April 1975, the Khmer Rouge seized power.

The Khmer Rouge, led by Pol Pot, dismantled all of Cambodian society. The government took control of all businesses and farms. It renamed the country Democratic Kampuchea. The Khmer Rouge regime forced most people in cities and towns to work on farms. Banks and currency were abolished, hospitals were closed, religion was almost completely abolished, education was strictly limited, and people had to dress alike. Large numbers of people were killed or died of mistreatment under the Khmer Rouge.

The Khmer Rouge remained in power until 1979, when invading Vietnamese and Cambodian allies seized control. Vietnam withdrew from Cambodia in 1989. In 1991, government and opposition groups signed a peace treaty. Multiparty elections were held in 1993, and a democratically elected government was established.

# CAMBODIA TODAY

Between 1975 and 1979, the Khmer Rouge killed large numbers of Cambodians, including many former government officials, intellectuals, and members of ethnic groups. In addition, conditions were so bad on the collective farms and work camps that many people died of exhaustion, starvation, and disease. Between 1 million and 3 million Cambodians may have died under the Khmer Rouge regime.

In January 1979, Vietnamese troops and Cambodian Communists ousted the Khmer Rouge from most of Cambodia. The new pro-Vietnamese leaders reduced government control of the economy and discontinued many Khmer Rouge policies. However, Cambodia had been shattered by years of warfare, and the country could no longer produce enough food to feed its people. Continued fighting from the Khmer Rouge and other groups presented severe obstacles to Cambodia's recovery.

In October 1991, Cambodia's government and opposition groups signed a United Nations-sponsored peace treaty. Under the treaty, the United Nations (UN) supervised the Cambodian government in 1992 through a transition to democracy. The UN worked with a 12-member Supreme National Council made up of members of the former government and the opposition groups.

A new constitution was put into effect in September 1993. A new democratically elected government was established, and it was headed by two prime ministers. In addition, the office of king was restored as a ceremonial position. Sihanouk, who had been head of state in a transitional government, became king. The Khmer Rouge boycotted the elections and did not join the new government.

Relations between Hun Sen and Prince Norodom Ranariddh, the two prime ministers, were strained. In July 1997, Hun Sen forced Ranariddh from office. Elections for the National Assembly were held in July 1998. Hun Sen's Cambodian People's Party won the most seats, and he remained prime minister. In 1999, Cambodia established a new Senate.

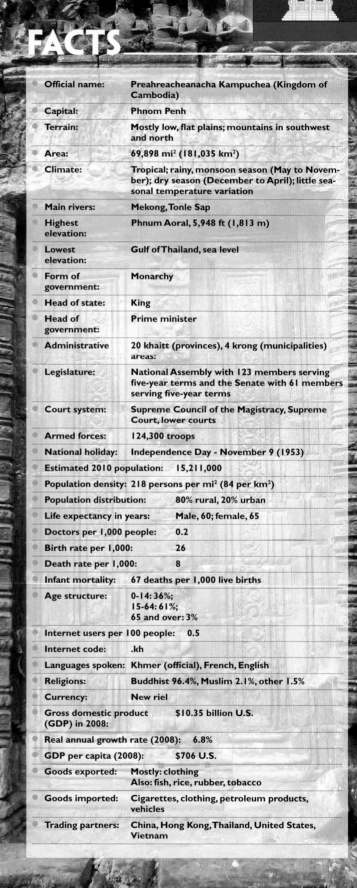

## FACTS

| | |
|---|---|
| Official name: | Preahreacheanacha Kampuchea (Kingdom of Cambodia) |
| Capital: | Phnom Penh |
| Terrain: | Mostly low, flat plains; mountains in southwest and north |
| Area: | 69,898 mi² (181,035 km²) |
| Climate: | Tropical; rainy, monsoon season (May to November); dry season (December to April); little seasonal temperature variation |
| Main rivers: | Mekong, Tonle Sap |
| Highest elevation: | Phnum Aoral, 5,948 ft (1,813 m) |
| Lowest elevation: | Gulf of Thailand, sea level |
| Form of government: | Monarchy |
| Head of state: | King |
| Head of government: | Prime minister |
| Administrative areas: | 20 khaitt (provinces), 4 krong (municipalities) |
| Legislature: | National Assembly with 123 members serving five-year terms and the Senate with 61 members serving five-year terms |
| Court system: | Supreme Council of the Magistracy, Supreme Court, lower courts |
| Armed forces: | 124,300 troops |
| National holiday: | Independence Day - November 9 (1953) |
| Estimated 2010 population: | 15,211,000 |
| Population density: | 218 persons per mi² (84 per km²) |
| Population distribution: | 80% rural, 20% urban |
| Life expectancy in years: | Male, 60; female, 65 |
| Doctors per 1,000 people: | 0.2 |
| Birth rate per 1,000: | 26 |
| Death rate per 1,000: | 8 |
| Infant mortality: | 67 deaths per 1,000 live births |
| Age structure: | 0-14: 36%; 15-64: 61%; 65 and over: 3% |
| Internet users per 100 people: | 0.5 |
| Internet code: | .kh |
| Languages spoken: | Khmer (official), French, English |
| Religions: | Buddhist 96.4%, Muslim 2.1%, other 1.5% |
| Currency: | New riel |
| Gross domestic product (GDP) in 2008: | $10.35 billion U.S. |
| Real annual growth rate (2008): | 6.8% |
| GDP per capita (2008): | $706 U.S. |
| Goods exported: | Mostly: clothing Also: fish, rice, rubber, tobacco |
| Goods imported: | Cigarettes, clothing, petroleum products, vehicles |
| Trading partners: | China, Hong Kong, Thailand, United States, Vietnam |

Phnom Penh, Cambodia's capital, lies on the Mekong River. This waterway serves as an important transportation network, since most of the country's roads and railways were destroyed during the conflicts of the 1970's and 1980's.

Cambodia is a small Southeast Asian country on the Gulf of Thailand. Widespread death and emigration during the 1970's and 1980's seriously reduced its population.

In National Assembly elections in 2003, the Cambodian People's Party (CPP) again won the most seats. However, the party did not win enough seats to govern alone. In 2004, the parties of Hun Sen and Norodom Ranariddh finally agreed to form a power-sharing government, with Hun Sen as prime minister. In 2008, the CPP won enough seats to govern alone, and Hun Sen remained prime minister.

By the late 1990's, most of the Khmer Rouge leaders had surrendered or were arrested. In 2004, Cambodia's National Assembly approved the creation of a special international court to try surviving Khmer Rouge leaders. Hearings began in 2009.

Also in 2004, King Sihanouk gave up the throne because of poor health. He was replaced by his son Norodom Sihamoni.

# ENVIRONMENT AND PEOPLE

Cambodia, with its warm, wet climate, flat land, and fertile soil, is ideal for growing rice. Since the 1970's, however, the country has struggled to pursue its traditional agricultural ways despite the devastating effects of warfare and the terror of the Khmer Rouge.

## Land and climate

Low mountains encircle most of Cambodia. Phnom Aoral, Cambodia's highest mountain peak, rises in the Cardamomes Range in the southwest. Fertile plains cover about a third of Cambodia's land, and forests cover the remainder.

The great Mekong River flows south from Laos through Cambodia. During the dry season, the Tonle Sap River flows southeast from the shallow *Tonle Sap* (Great Lake) and joins the Mekong at Phnom Penh. During the *monsoon* (rainy) season, the river flows in the opposite direction because floods and melted snow from the Mekong's source in Tibet raise the river waters to a level higher than that of the lake.

Cambodia has a tropical climate. At Phnom Penh, the daily temperature averages about 85 °F (29 °C) all year round. During the rainy season, which lasts from May to November, annual rainfall ranges from less than 60 inches (150 centimeters) in Phnom Penh to about 200 inches (510 centimeters) on the coast.

The Mekong River is the largest waterway on the Southeast Asian peninsula. As it flows through Cambodia, the Mekong waters many rice fields. In the rainy season, the river sometimes floods the flat, low-lying Cambodian landscape.

## A transformed society

Years of warfare and Khmer Rouge rule shattered the Cambodian economy as well as many of its cherished traditions. However, since the Khmer Rouge regime was overthrown in 1979, Cambodians have begun to rebuild their society. But the principles of Communism still affect Cambodian life.

Before the Khmer Rouge took over, most Cambodians were rice farmers, and rural families owned the land they lived on and farmed. However, the Khmer Rouge abolished private ownership of land and forced people to work on communal farms, where even rice for meals was often strictly rationed. Cultivation was greatly disrupted, and the country suffered grave food shortages.

Today, as before, most Cambodians are rice farmers, living in villages of between 100 and 400 people and

A worker in a textile factory manufactures cotton cloth. Cambodian factories produce plastic, tools, textiles, cigarettes, and other consumer goods. However, the development of industry is slowed by lack of power and raw materials.

After the Vietnamese-backed government took over, education made a slow recovery. However, it is still limited by lack of school buildings, teachers, and books and supplies. Children are required to attend school from the ages of 6 to 12. In 1975, the Khmer Rouge closed Phnom Penh University. But teaching in the fields of medicine and pharmacy resumed in 1980, and the university officially reopened in 1988.

Like education, religion also was violently suppressed under the Khmer Rouge. Many monks were killed, and others were taken from the monasteries and forced to perform hard labor. Temples were destroyed or converted to nonreligious uses. Today, however, the ban on Buddhism has been lifted. Although the government still limited religious practice in the 1980's, Buddhism was reinstated as the national religion in 1989.

working on paddies near their villages. Like the Khmer Rouge, the present government also prohibits private ownership of land. But some rural Cambodians may keep the rice they grow, and a few are allowed to occupy the land they owned before the Khmer take-over. Although the agricultural situation has improved somewhat, Cambodia still has serious problems with food production.

From the early 1900's until 1975, public education was widespread in Cambodia, and students learned a variety of subjects, including arithmetic, history, geography, health, language, and science. University education was also available. After 1975, the Khmer Rouge killed thousands of teachers and almost completely destroyed the educational system.

# ANGKOR

Angkor was an early Khmer civilization that flourished in northwestern Cambodia from the early 800's to the 1400's. The most famous capital of this civilization also was called Angkor. *Angkor* means *city,* or *capital.*

The Khmer built a huge imperial capital at Angkor. It may have had a million people, more than any European city at that time. Angkor included Angkor Thom, a "city within a city" that covered 4 square miles (10 square kilometers); Angkor Wat, a grand Hindu temple; and many other temples and palaces.

A number of building materials were used at Angkor, including brick, laterite, and sandstone. Laterite, a type of stone, provided material for foundations. Doorways, towers, and decorative details were built of sandstone, often carved in beautiful, complex designs.

Angkor Thom's central temple, the Bayon, an impressive structure with more than 200 giant stone faces adorning its towers, was erected in honor of Buddha and the reigning king. However, Angkor Wat was the most magnificent temple in Angkor.

## Angkor Wat

Built in the early 1100's and dedicated to the Hindu god Vishnu, Angkor Wat was used as an astronomical observatory, as well as for religious purposes. It later became the tomb of the Cambodian king who had ordered its construction.

The vast scale of Angkor Wat reflects the enormous wealth of the Khmer empire. The monument covers nearly 1 square mile (2.6 square kilometers) and is surrounded by a moat.

Angkor Wat consists of a group of temples constructed in a pyramidal form that imitates

the mythological home of the Hindu gods. A series of staircases and terraces lead to the highest tower. The galleries that surround the central pyramid are decorated with carved scenes of Khmer history, as well as images from Hindu mythology.

## The empire's decline

Many new temples and other buildings were built in the Angkor kingdom in the late 1100's and early 1200's. However, the great speed at which the temples were built often led to careless workmanship and faster deteriora-

Sculptures of demons in the Hindu religion stand in the temple complex of Angkor Wat.

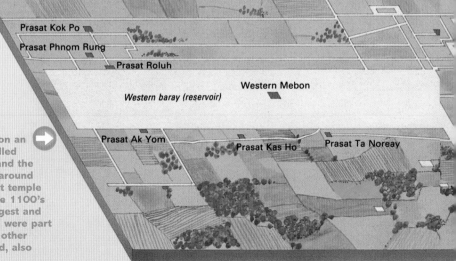

Prasat Kok Po
Prasat Phnom Rung
Prasat Roluh
Western Mebon
Western baray (reservoir)
Prasat Ak Yom
Prasat Kas Ho
Prasat Ta Noreay

The ancient Khmer capital of Angkor was built upon an elaborate network of canals and artificial lakes called barays. The canals irrigated the city's rice fields, and the barays served as reservoirs. The Bayon (A), built around 1200 and dedicated to Buddha, was the last great temple built within the capital. Angkor Wat (B), built in the 1100's and dedicated to the Hindu god Vishnu, is the largest and best preserved of the city's temples. Moats, which were part of the water system, surrounded Angkor Wat and other temples. Angkor Thom, where the Bayon is located, also was surrounded by a moat.

tion. From that time to the end of the empire, little new construction was undertaken.

This deterioration perhaps symbolized the fate of the Angkor empire, which began to decline after the 1100's. Invasions from Thailand, epidemics of malaria, and disputes within the royal Cambodian family may have caused the empire's downfall. Thai forces captured the capital in 1431 but soon abandoned it. Looters over-ran the temples, and years of forest growth covered the once-beautiful city.

Angkor remained abandoned for more than 400 years. However, interest in Cambodia's "lost city" was stirred by accounts of European explorers written in the 1600's. Early maps of the region eventually led to the ancient Khmer capital's rediscovery in 1860 by Henri Mouhot, a French naturalist.

Mouhot uncovered stone carvings and manuscripts that revealed much about daily life in Angkor. One manuscript described the magnificence and riches of one of Angkor's temples and noted that thousands of people were required to maintain it.

From the 1860's to the mid-1900's, French and Cambodian archaeologists restored and rebuilt many of Angkor's temples. A great deal of damage had been done by tree roots that had pushed up under the foundations, and rain had destroyed many of the exquisite sandstone carvings and statues. Nevertheless, even in their ruined state, the temples of Angkor probably represent the finest architectural monuments in Cambodia.

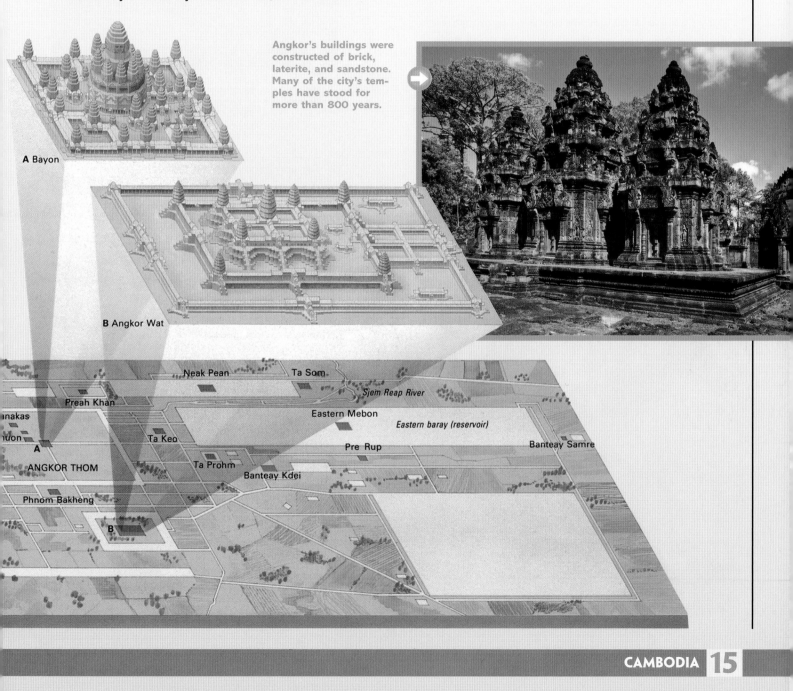

Angkor's buildings were constructed of brick, laterite, and sandstone. Many of the city's temples have stood for more than 800 years.

A Bayon

B Angkor Wat

Neak Pean  Ta Som
Preah Khan  Siem Reap River
anakas  Eastern Mebon
uon  Eastern baray (reservoir)
A  Ta Keo
ANGKOR THOM  Pre Rup  Banteay Samre
Ta Prohm
Banteay Kdei
Phnom Bakheng
B

# CAMEROON

The west African country of Cameroon is home to more than 19 million people from about 200 different ethnic groups. The largest groups are the Bamiléké, who live in the western mountains of the country, and the Fulani, who live in the north. The Douala, the Ewondo, and the Fang inhabit the southern and central sections.

English and French are the official languages of Cameroon, holdovers from the time when the country was a colony of both the United Kingdom and France. But most Cameroonians speak one of the country's many African languages.

Many of these languages are Bantu—a term that refers to both the languages and the people who speak them. About 300 Bantu languages and dialects are spoken in Africa.

Several hundred years before Christ was born, the first Bantu people lived in the northern highlands of what is now Cameroon. About the time of Christ, the Bantu began one of the greatest migrations in history.

Traditional clay huts have sheltered northern Cameroonians for hundreds of years. Many people in this hot, dry region belong to the Fulani ethnic group, one in Cameroon.

The migration occurred gradually, with small groups continually splitting off and moving to new regions as their population expanded. The groups eventually formed about 300 separate ethnic groups, and by the year 1500, Bantu peoples were living in most of central, eastern, and southern Africa. The Bassa, Douala, Ewondo, and Fang are among the Bantu peoples living in Cameroon today.

Cameroon's Pygmies are a small but special group of people, probably one of the first groups to live in Africa. Pygmies are only about 4 feet to 4 feet 8 inches (1.2 to 1.42 meters) tall, with reddish-brown skin and tightly curled brown hair.

The Pygmies' home has always been the thick tropical rain forests of central Africa, where they live in small bands, hunting and gathering. But today the Pygmies of Cameroon have a much smaller area to live in. Over hundreds of years, Bantu-speaking peoples invaded much of the Pygmies' territory and cut down the forests. The Pygmies lost even more land as people

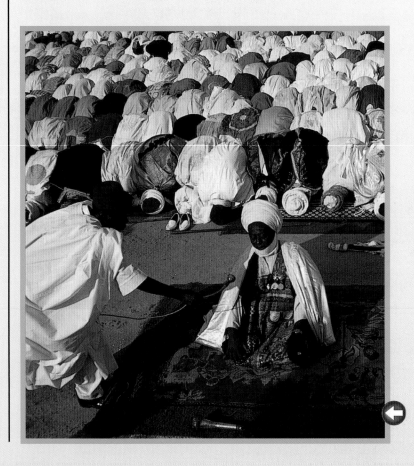

A Muslim religious leader leads worshipers in prayer. About 20 percent of Cameroon's people are Muslims, and about 40 percent follow traditional African religions.

built roads and towns in the forests. Fewer and fewer Pygmies in Cameroon and neighboring countries are able to follow their traditional way of life.

Most of the other people in Cameroon also live off the land, much as their ancestors did. Nearly half of the people make their homes in rural areas, mainly in villages or small towns, and many make their living as farmers or herders.

Houses in the northern towns and villages are round clay huts or rectangular brick houses. Herders who roam the north build light shelters from poles and woven mats so that they can move easily from place to place. Houses in the western mountains are usually square brick structures, while homes in the central and southern forests are typically made of wood, palm leaves, and clay. Along the coast, people build wooden houses and cover them with tree bark or sheets of metal. In the cities, some Cameroonians live in modern houses and apartments, but others live in rundown shacks in slum conditions.

The Cameroon government offers free public education, and private schools are given financial aid. However, due to a severe shortage of schools and teachers, many children do not attend school at all. Among Cameroonians 15 years or older, more than half can read and write.

Colorful robes and a golden nose ring are part of the traditional dress of a woman in northern Cameroon.

Traders in a street market, conduct their business in the shade of leafy trees. Village markets are often social centers where many rural Cameroonians gather to exchange news and visit with friends.

The slave trade became a triangular route across the Atlantic Ocean after the establishment of European plantations in the Americas during the 1500's. Food and manufactured goods were loaded onto cargo ships in such European ports as Liverpool, England, and Lisbon, Portugal. The ships sailed to the African coast, where they sold their cargoes and took on a new load—African slaves. Then they sailed across the ocean to the Western Hemisphere, where the slaves were sold to work on plantations. The ships then took on cargoes of plantation products, such as tobacco, sugar, and cotton, for transport back to Europe.

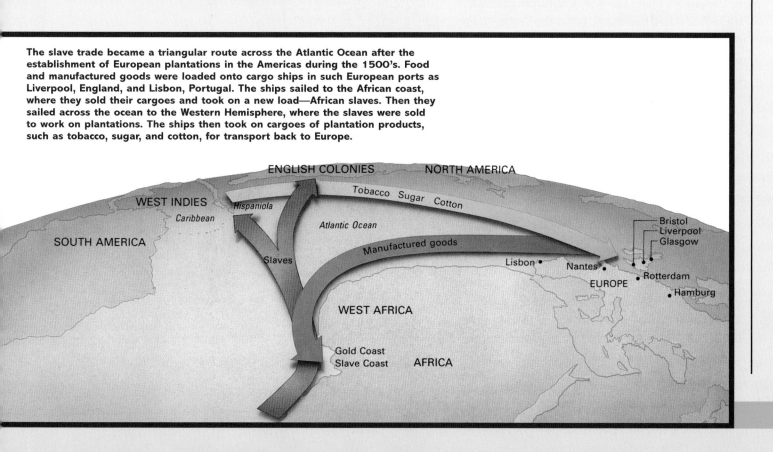

ENGLISH COLONIES    NORTH AMERICA

WEST INDIES    Hispaniola
Caribbean    Atlantic Ocean
SOUTH AMERICA
Tobacco  Sugar  Cotton
Bristol
Liverpool
Glasgow
Manufactured goods
Slaves
Lisbon    Nantes
Rotterdam
EUROPE
Hamburg
WEST AFRICA
Gold Coast
Slave Coast    AFRICA

# CAMEROON TODAY

Like many African countries, modern Cameroon is a blend of traditional ways and modern challenges. In rural areas, Cameroonians live much as their ancestors did, while in the cities the people face such problems as poverty and overcrowding as they work to develop Cameroon into a contemporary society.

Cameroon is a republic. A president is Cameroon's head of state. This official holds the most power in the government. A 180-member National Assembly makes the country's laws. The president appoints the prime minister, Cabinet members, and other officials to help carry out the functions of the government. The people elect both the president and the Assembly.

## Historical background

Portuguese explorers, who arrived in the 1400's, were the first Europeans to see Cameroon. They found huge schools of shrimplike animals, which they called camarões (shrimp), in the Wouri River. That Portuguese word eventually led to the country's name.

For the next 300 years, other Europeans flocked to the region. Many came to profit from the slave trade that flourished there. When the United Kingdom outlawed its slave trade in 1807—followed by other European nations in the early 1800's—traders turned to ivory and palm oil.

During the 1800's, three European countries—the United Kingdom, France, and Germany—competed for control of Cameroon. In 1858, British missionaries established Victoria, Cameroon's first permanent European settlement, at the base of Mount Cameroon. In 1884, however, two local chiefs of the Douala people in Cameroon signed a treaty with Germany that made the region a German protectorate.

Then, during World War I (1914-1918), Germany lost control of Cameroon to the United Kingdom and France, and in 1922 those two countries divided the land between them. The British section, called the British Cameroons, consisted of two separate parts along the western border. The French section, called

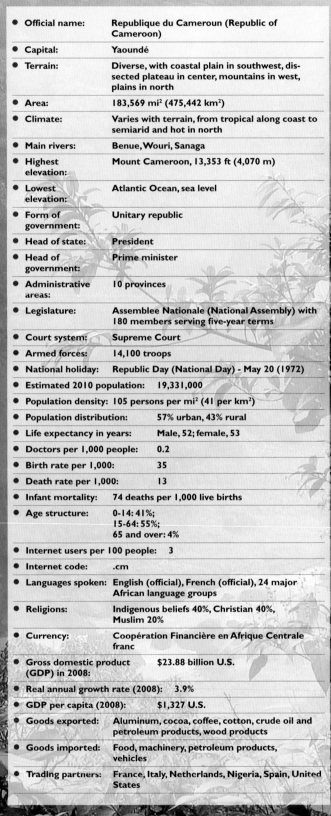

## FACTS

| | |
|---|---|
| Official name: | Republique du Cameroun (Republic of Cameroon) |
| Capital: | Yaoundé |
| Terrain: | Diverse, with coastal plain in southwest, dissected plateau in center, mountains in west, plains in north |
| Area: | 183,569 mi² (475,442 km²) |
| Climate: | Varies with terrain, from tropical along coast to semiarid and hot in north |
| Main rivers: | Benue, Wouri, Sanaga |
| Highest elevation: | Mount Cameroon, 13,353 ft (4,070 m) |
| Lowest elevation: | Atlantic Ocean, sea level |
| Form of government: | Unitary republic |
| Head of state: | President |
| Head of government: | Prime minister |
| Administrative areas: | 10 provinces |
| Legislature: | Assemblee Nationale (National Assembly) with 180 members serving five-year terms |
| Court system: | Supreme Court |
| Armed forces: | 14,100 troops |
| National holiday: | Republic Day (National Day) - May 20 (1972) |
| Estimated 2010 population: | 19,331,000 |
| Population density: | 105 persons per mi² (41 per km²) |
| Population distribution: | 57% urban, 43% rural |
| Life expectancy in years: | Male, 52; female, 53 |
| Doctors per 1,000 people: | 0.2 |
| Birth rate per 1,000: | 35 |
| Death rate per 1,000: | 13 |
| Infant mortality: | 74 deaths per 1,000 live births |
| Age structure: | 0-14: 41%; 15-64: 55%; 65 and over: 4% |
| Internet users per 100 people: | 3 |
| Internet code: | .cm |
| Languages spoken: | English (official), French (official), 24 major African language groups |
| Religions: | Indigenous beliefs 40%, Christian 40%, Muslim 20% |
| Currency: | Coopération Financière en Afrique Centrale franc |
| Gross domestic product (GDP) in 2008: | $23.88 billion U.S. |
| Real annual growth rate (2008): | 3.9% |
| GDP per capita (2008): | $1,327 U.S. |
| Goods exported: | Aluminum, cocoa, coffee, cotton, crude oil and petroleum products, wood products |
| Goods imported: | Food, machinery, petroleum products, vehicles |
| Trading partners: | France, Italy, Netherlands, Nigeria, Spain, United States |

↑
Yaoundé is the capital of Cameroon. The city also is the country's commercial and transportation center.

Cameroon is a country in western Africa, where the continent's Atlantic coastline turns from an east-west direction to north-south.

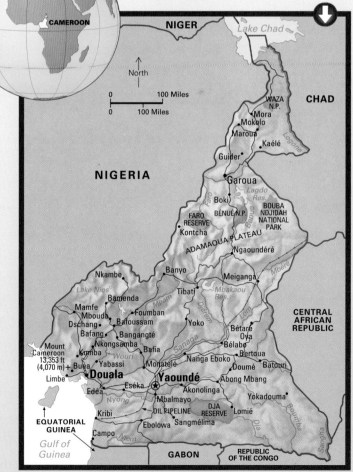

French Cameroon, consisted of the remaining land—about 80 percent of present-day Cameroon. Each section was governed by the laws of its ruling country and adopted that country's language.

## Independence

On Jan. 1, 1960, French Cameroon became the independent Republic of Cameroon. In an election in 1961, the United Kingdom asked the people in the British Cameroons whether they wanted to join the new republic or unite with neighboring Nigeria. Voters in the northern part chose to join Nigeria and became part of that country on June 1. Voters in the southern part chose to join Cameroon, and that union took place on Oct. 1. Ahmadou Ahidjo became president.

From 1961 until early 1972, Cameroon operated as a federation of two states—East Cameroon and West Cameroon. In May 1972, Cameroon adopted a new constitution that united the two states.

Ahidjo retired in 1982, and Paul Biya succeeded him. Until 1991, Cameroon only allowed one political party. But late that year, the government legalized opposition parties. Biya won reelection in 1992, 1997, 2004, and 2011 in multiparty elections that many observers considered to be flawed.

# LAND AND ECONOMY

Cameroon covers more than 180,000 square miles (470,000 square kilometers) of mountains, grasslands, and forests in west-central Africa. The mountains and hills run along Cameroon's western border, from the seacoast in the south to Lake Chad in the north. Near the coast, Mount Cameroon—Cameroon's highest point—rises 13,353 feet (4,070 meters) above sea level.

A grassy, thinly wooded plain, or savanna, in northern Cameroon contains a wildlife reserve called Waza National Park, the home of a great variety of African animals, including elephants, giraffes, monkeys, and antelopes. This northern savanna region is hot and dry most of the year. The average temperature is about 82° F (28° C), but daytime temperatures sometimes reach as high as 120° F (49° C).

A forested plateau covers central Cameroon. The central plateau is cooler than the savanna, with an average temperature of about 75° F (24° C).

South of the plateau, a tropical forested lowland lies along the coast. This coastal region is hot and humid, with an average temperature of about 80° F (27° C). Some areas receive up to 200 inches (500 centimeters) of rain each year.

Huge numbers of wild animals once roamed the forests of southern Cameroon, but they have become scarce. Many were killed by hunters, and others by farmers trying to protect their crops.

## A developing economy

Although Cameroon is a developing country, and quite poor by Western standards, its economy is strong and varied compared to other countries in western Africa.

Cameroon's economy still depends heavily on agriculture, and most of the nation's people are farmers or herders. Farmers grow such food crops as cassava, corn, millet, yams, and sweet potatoes, mainly for their own use. They also raise such cash crops as bananas, cacao beans, coffee, cotton, and peanuts.

Petroleum is Cameroon's most important natural resource. Since the 1970's, when the country began producing petroleum from wells drilled in the Gulf of Guinea, oil exports have greatly aided the economy. Together with hydroelectric power, the petroleum has also helped the country meet its energy needs.

Carrying their burdens on their heads in traditional style, two women walk along a hot, dusty road in Cameroon's northern savanna region. There, daytime temperatures may reach 120° F (49° C), and little rain falls.

Passengers reach dry land after a journey across the Sanaga River in southern Cameroon. Some of the energy needed for the country's developing industries comes from hydroelectric projects on rivers such as the Sanaga.

Cameroon's other mineral products include limestone and *pozzuolana* (a rock used to make cement). The country also has deposits of bauxite, iron ore, diamonds, and other minerals, but these are not mined commercially. Cameroon's natural resources also include its vast forests, which provide palm oil, rubber, and timber.

Most of Cameroon's manufacturing industries are based on the processing of the nation's resources. Some plants, for example, process agricultural raw materials, such as animal hides, cacao, and sugar cane. Petroleum refining and the manufacture of cement from limestone and pozzuolana are also important. Other factories manufacture products using raw materials or parts imported from other countries. For example, aluminum is produced from bauxite imported from nearby Guinea. Other factories in Cameroon manufacture consumer products such as beer, cigarettes, shoes, soap, and soft drinks.

Rural villages or small towns are home to about two-thirds of Cameroon's people. Villagers raise food crops for their families or tend livestock. Cameroon is still a developing nation.

## LAKE NIOS

On Aug. 21, 1986, a cloud of carbon dioxide gas escaped from the waters of Lake Nios in rural western Cameroon. This colorless, odorless gas, normally present in very small amounts in the atmosphere, is heavier than oxygen. As the cloud of gas left the lake, it flowed downhill toward a number of small villages, displacing the air as it went. As a result, more than 1,700 Cameroonians, unable to breathe oxygen, died of suffocation. The bodies of their cattle lay scattered across the fields, and even birds and insects were killed. The gas apparently had been dissolved in water deep in the lake, but no one knows what caused the gas to be released. Witnesses reported hearing a low rumble right before the disaster, and some scientists think that an earth tremor or some kind of volcanic activity beneath the lake triggered the violent release of carbon dioxide.

Lake Nios lies in the mountainous region of western Cameroon, near the Nigerian border.

## City life

Manufacturing industries and some service industries provide jobs in Cameroon's urban areas, where more than half of the people live. Each year, large numbers of rural people move to the towns and cities to find jobs, causing problems with overcrowding and housing.

Yaoundé and Douala, the largest cities, have elegant hotels, fine office buildings, and fashionable modern houses. But they also have slum areas where many poorer people live in wretched conditions.

Little industry is located in Yaoundé, the capital of Cameroon, but the city is a busy commercial and transportation center. Railroads connect it with Douala, the country's largest city and the Gulf of Guinea's major port. Road transportation is difficult in Cameroon because most roads are unpaved.

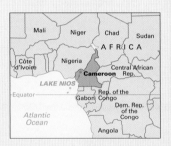

Lake Nios • Cha • Bu

• Nios

Djingbe •

• Subum

Gas cloud shown in yellow

# CANADA

Canada is a land of many contrasts. It is the second largest country in the world, but it ranks thirty-sixth in population, with about 33 million people. It is slightly larger than the United States, but has only about one-tenth as many people.

Canada spans the continent of North America from Newfoundland and Labrador on the Atlantic Ocean to British Columbia on the Pacific, and from the Arctic Ocean in the north to the United States in the south. Yet 75 percent of the Canadian people live within 100 miles (150 kilometers) of the country's southern border. Few people live in the rest of the country because of its rugged terrain and harsh climate.

The Canadian landscape includes towering mountains in the west; vast, flat prairies inland; and sandy beaches on the eastern coast. In the north, lush forests cover huge areas of land, and beyond the woodlands lies the frozen, barren Arctic.

Canada is wealthy in natural resources. Beginning in the 1500's, Europeans came to fish in its coastal waters and trap fur-bearing animals in its forests. The forests also became a resource as settlers used their wood for building and fuel. Fertile soil attracted immigrant farmers. Today, Canada's mighty rivers supply water and power, and the nation's deposits of petroleum and other minerals provide raw materials for manufacturers.

Canada's people are as varied as its landscape. Most Canadians have ancestors who are of English, French, Irish, or Scottish descent. Both English and French are official languages today. French Canadians, most of whom live in the province of Quebec, have kept the language and many customs of their ancestors. Other large ethnic groups include Chinese, German, Italian, and Ukrainian people. Large numbers of Asians live in western Canada and Ontario. Native peoples—American Indians and Inuit—make up a small percentage of the nation's population.

A large majority of Canada's people live in cities and towns. Many of these urban areas lie near the Great Lakes and the St. Lawrence River, including Montreal and Toronto, the largest cities in Canada. Inuit and American Indians make up about 40 percent of the population of the Arctic north.

Canada is an independent, self-governing nation with strong historic ties to the United Kingdom. It has a close and friendly relationship with the United States but strives to keep its Canadian identity. The name *Canada* probably came from an Iroquois word meaning *community,* though keeping a sense of community over such a vast area continues to challenge the Canadian people.

**23**

# CANADA TODAY

Canada is a member of the Commonwealth of Nations, an association of countries once governed by the United Kingdom. Today, it is an independent democracy with a federal form of government like the United States, combined with a cabinet system and Parliament like the United Kingdom.

The national government in Ottawa takes care of matters that affect the nation as a whole. Canada is made up of 10 provinces and 3 territories. Each province has its own government; the national government helps administer the territories.

Because Canada is a member of the Commonwealth, the queen of England is also queen of Canada, but a governor general acts as her representative. The governor general once had great power but now performs only formal and symbolic duties. The prime minister heads the government.

The cabinet system of Canada combines the legislative and executive branches of government. Parliament is the national legislature. It has two houses: the House of Commons and the Senate.

Members of the House are elected by the people to four-year terms unless a new election is called earlier. The prime minister is the leader of the majority party in the House of Commons. The prime minister recommends people for the Senate, and they are appointed by the governor general.

The office of prime minister is not established by law. It became a custom long ago when people saw that a leader was needed in this kind of government. The prime minister chooses as many as 40 ministers to head departments in the Cabinet. The prime minister and the Cabinet run the government with the support of Parliament.

Parliament can control the actions of the prime minister by withholding support. If this happens, the prime minister must resign or ask the governor general to call a new election. In turn, the prime minister can control the actions of Parliament by asking for a new election.

## FACTS

| | |
|---|---|
| Official name: | Canada |
| Capital: | Ottawa |
| Terrain: | Mostly plains with mountains in west and lowlands in southeast |
| Area: | 3,855,103 mi² (9,984,670 km²) |
| Climate: | Varies from temperate in south to subarctic and arctic in north |
| Main rivers: | St. Lawrence, Fraser, Mackenzie, Nelson |
| Highest elevation: | Mount Logan, 19,551 ft (5,959 m) |
| Lowest elevation: | Atlantic Ocean, sea level |
| Form of government: | Constitutional monarchy |
| Head of state: | British monarch, represented by governor general |
| Head of government: | Prime minister |
| Administrative areas: | 10 provinces, 3 territories |
| Legislature: | Parliament or Parlement consisting of the Senate or Senat with 105 members serving until age 75 and the House of Commons or Chambre des Communes with 308 members serving five-year terms |
| Court system: | Supreme Court |
| Armed forces: | 64,400 troops |
| National holiday: | Canada Day - July 1 (1867) |
| Estimated 2010 population: | 33,772,000 |
| Population density: | 9 persons per mi² (3 per km²) |
| Population distribution: | 81% urban, 19% rural |
| Life expectancy in years: | Male, 78; female, 83 |
| Doctors per 1,000 people: | 1.9 |
| Birth rate per 1,000: | 11 |
| Death rate per 1,000: | 7 |
| Infant mortality: | 5 deaths per 1,000 live births |
| Age structure: | 0-14: 17%; 15-64: 69%; 65 and over: 14% |
| Internet users per 100 people: | 73 |
| Internet code: | .ca |
| Languages spoken: | English (official), French (official) |
| Religions: | Roman Catholic 43.6%, Protestant 29.2%, other Christian 4.3%, Muslim 2%, other 20.9% |
| Currency: | Canadian dollar |
| Gross domestic product (GDP) in 2008: | $1.511 trillion U.S. |
| Real annual growth rate (2008): | 0.6% |
| GDP per capita (2008): | $45,699 U.S. |
| Goods exported: | Motor vehicles and parts, natural gas, petroleum, wheat, wood products |
| Goods imported: | Computers, machinery, motor vehicles and parts, pharmaceuticals, scientific equipment |
| Trading partners: | Mostly: United States Also: China, France, Germany, Japan, Mexico, South Korea, United Kingdom |

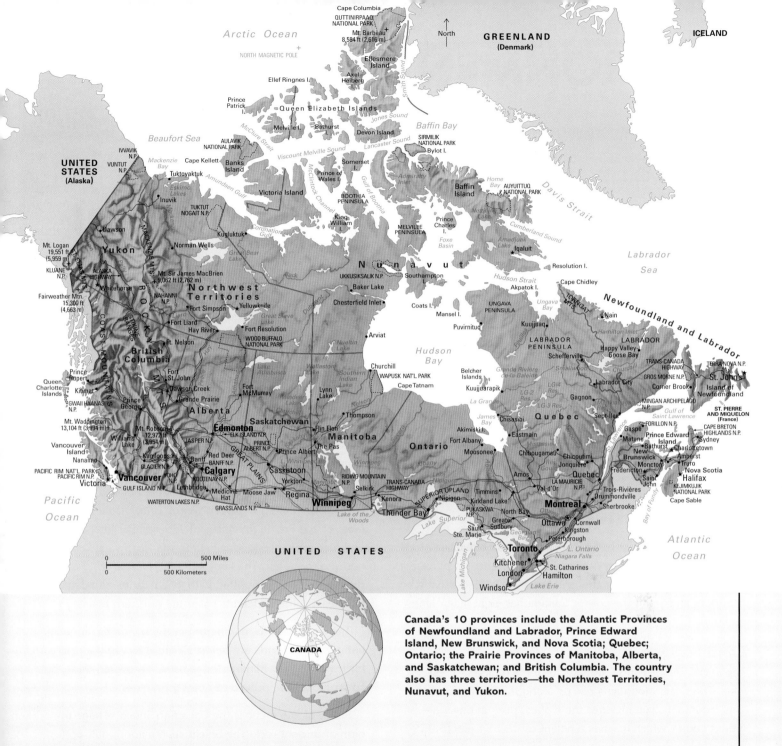

Canada's 10 provinces include the Atlantic Provinces of Newfoundland and Labrador, Prince Edward Island, New Brunswick, and Nova Scotia; Quebec; Ontario; the Prairie Provinces of Manitoba, Alberta, and Saskatchewan; and British Columbia. The country also has three territories—the Northwest Territories, Nunavut, and Yukon.

The Supreme Court of Canada is Canada's highest court. There are also federal courts, provincial and territorial courts, the Tax Court of Canada, and *martial* (military) courts.

Canada's government is based on a constitution that is partly unwritten and partly written. The unwritten part is based mainly on custom, such as the cabinet system and the office of prime minister. The written part consists of the Constitution Act of 1982, the British North America Act, and various laws and court decisions. The

Constitution Act of 1982 ended formal British control over amendments to Canada's constitution.

The British North America Act, which governed Canada from 1867 to 1982, established a powerful national government. Through the years, however, the provinces became stronger and richer. Arguments about the division of power between federal government and provincial governments increased. Each province now controls such matters as education, administration of justice, property, and civil rights.

# ENVIRONMENT

Banff National Park is famous for its spectacular scenery, including the snow-capped Canadian Rockies.

The Horseshoe Falls stand on the Canadian side of Niagara Falls.

Canada's vast and varied land covers most of the northern half of North America, extending 3,223 miles (5,187 kilometers) from east to west, over six time zones. It stretches 2,875 miles (4,627 kilometers) from north to south. Only Greenland lies nearer to the North Pole.

Forests cover almost half of Canada's land area, while mountains and the frozen Arctic make up another 41 percent. Only about 12 percent of the land has been settled, mostly in the south and along the coasts. Canada has eight major land regions.

## Land regions

The Pacific Ranges and Lowlands form the region farthest west. The Coast Mountains rise up along the Pacific coast. This mountain chain includes the highest peak in Canada—Mount Logan, which reaches a height of 19,551 feet (5,959 meters) in Yukon.

The coast of British Columbia is cut by *fiords* (long, narrow inlets) that reach the dense forests on the lower slopes of the mountains. The tall western red cedars, Douglas firs, hemlocks, and other evergreens here are a valuable Canadian resource. East of the Coast Mountains is a mineral-rich area of grassy plains, fertile valleys, and smaller mountains.

The Rocky Mountains form the region to the east of the Pacific Ranges and Lowlands. The snow-capped peaks and sky blue lakes of the Rockies offer some of the world's most beautiful scenery. Thousands of tourists visit the mountains of British Columbia and western Alberta every year.

The Interior Plains stretch east of the Rockies. This region covers the northeast corner of British Columbia and most of the Prairie Provinces of Alberta, Saskatchewan, and Manitoba. It extends north to the Northwest Territories and the Arctic. The southern plains, once vast natural grasslands, are now dotted with wheat farms and cattle ranches. Evergreen forests cover the northern plains.

The Arctic Islands, numbering in the hundreds, are uninhabited. The islands are *tundras*—areas too cold and dry for trees to grow. Minerals have been discovered there, but so far most have proved too costly to mine.

South of the Arctic is the Canadian Shield, a huge region of ancient rock. The Shield covers about half of Canada and curves in a horseshoe shape around Hudson Bay. The low hills of the Canadian Shield are actually very old mountains worn down by glaciers. The glaciers also scraped away most of the soil and left behind thousands of lakes, which are the source of many rivers. Valuable evergreen forests cover much of the Canadian Shield, and the region is rich in minerals as well.

The Hudson Bay Lowlands are a flat, swampy region extending southwest from Hudson Bay. Few people live in this area of forests and vast deposits of *peat* (the decayed remains of plants).

North Pole
Arctic Ocean
NORTH MAGNETIC POLE
Ellesmere Island
Beaufort Sea
Queen Elizabeth Islands
ARCTIC ISLANDS
Baffin Bay
Banks I.
Victoria I.
Boothia Peninsula
Baffin Island
Melville Peninsula
Davis Strait
Arctic Circle
Foxe Basin
Labrador Sea
Great Bear Lake
Southampton
Cape Chidley
Hudson Strait
Mt. Logan 19,551 ft (5,959 m)
Yukon Plateau
Yukon
Mackenzie
Back
Great Slave Lake
Dubawnt
Ungava Peninsula
ROCKY MOUNTAINS
Coast Mountains
CANADIAN SHIELD
Hudson Bay
Labrador Peninsula
Island of Newfoundland
Peace
L. Athabasca
Churchill
Reindeer Lake
Queen Charlotte Islands
Nelson
HUDSON BAY LOWLANDS
CANADIAN SHIELD
La Grande
Cape Race
PACIFIC RANGES AND LOWLANDS
Athabasca
Edmonton
Saskatchewan
APPALACHIAN HIGHLANDS
Laurentian Mts.
Cape Breton I.
Fraser
INTERIOR PLAINS
Lake Winnipeg
Albany
James Bay
St. Lawrence
Vancouver I.
Vancouver
Great Plains
L. Manitoba
Winnipeg
Northern Highland
Ottawa
Montreal
Bay of Fundy
North Pacific Ocean
L. Superior
L. Huron
Toronto
L. Ontario
North Atlantic Ocean
ST. LAWRENCE LOWLANDS
L. Michigan
L. Erie

**Canada spreads over most of the northern half of North America. It covers 3,855,103 square miles (9,984,670 square kilometers), including inland water such as Great Bear Lake, Great Slave Lake, and thousands of other lakes. The country also has one of the longest coastlines of any country. Hudson Bay and James Bay form a huge inland sea. Hudson Bay, which remains frozen for about eight months of the year, serves as a summer waterway to Canada's vast interior. The Coast Ranges and Rocky Mountains in western Canada are sometimes called the Cordillera. A great northern forest sweeps across Canada from Alaska to Newfoundland and Labrador. Near the Arctic Ocean, the forests gradually give way to treeless, mossy tundra that lies over frozen soil called permafrost. Crops are grown mainly on the southern Interior Plains and in the St. Lawrence Lowlands.**

The St. Lawrence Lowlands, Canada's smallest land region, is home to more than half of all Canadians. The area includes the St. Lawrence River Valley and southern Ontario, where crops thrive on the flat or gently rolling land. Canada's major cities of Quebec, Montreal, and Toronto are located here. The region is Canada's manufacturing center.

The Appalachian Region includes southeastern Quebec as well as all the Atlantic Provinces except Newfoundland and Labrador. The rounded Appalachian Mountains dominate this region. Most of the people live along the Atlantic coast, where hundreds of inlets provide excellent harbors for fishing fleets.

**Forests line a scenic lake near Haliburton, Ontario.**

## Climate

Canada's northern location gives the country a generally cold climate with average January temperatures below 0 °F (-18 °C) in more than two-thirds of the country. Northern Canada has short, cool summers, but in southern Canada, summers are long enough and warm enough for raising crops. Precipitation is heaviest along the Pacific coast, where lush forests grow. Southeastern Canada also receives plentiful precipitation, including heavy snow in the winter months.

**A birdhouse hangs on a fence in a vast Alberta prairie.**

# THE FIRST CANADIANS

Canada's history begins with the Asiatic peoples who crossed into what is now North America at least 15,000 years ago. These people moved southward and became the first North Americans. In Canada today, they are known as "Indians," "First Nations," or "native peoples."

The ancestors of the Inuit also came from Asia, probably about 5,000 years ago, and settled in the Arctic area. *Inuit* means *the people* in the Inuit-Inupiaq language. Some people call them *Eskimos,* which comes from an Indian word meaning *eaters of raw meat.*

## Traditional ways

People living in different parts of Canada eventually developed different ways of life. The Inuit and Indian peoples of Canada have been grouped into six main culture areas, determined by their traditional ways of life—the Arctic, Subarctic, Northwest Coast, Plateau, Plains, and Northeast (also called Eastern Woodlands) areas.

The Inuit had to survive in the harsh environment of the Arctic. The sea provided most of their food—seals, walruses, whales, and fish. The Inuit also hunted polar bears and caribou. They used animal skins to make clothing, summer tents, and boats. In the winter, they made sod houses or snowhouses, and they traveled by dog sled.

Most Indians of the vast Subarctic region of northern Canada belonged to two major language groups: the Athabaskan in the west and the Algonquian in the east. These people lived mainly by hunting, gathering, and fishing. They used wood from the northern forests to make utensils, houses with wooden frames, bark canoes, and snowshoes.

Native peoples of the Northeast lived in southeastern Canada. Groups in the northern part of the area lived by hunting, gathering, and fishing. Farther south, they also farmed. By the 1600's, important groups included the Huron and the tribes of the Iroquois League.

Prairie tribes, such as the Blackfeet and the Assiniboine, lived on Canada's southern prairies.

Beautiful Dancing Blankets, such as this one are used by the Tlingits of British Columbia. The blankets are used as ceremonial robes and as the dancers move, the fringe provides an exciting image. The blankets' abstract patterns represent clan symbols or forms from nature.

Northwest Coast Indians created totem poles (1), dugout canoes (2), rich dress (3), and masks (4). Inuit used harpoons (5) to hunt.

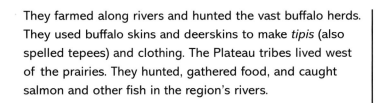

They farmed along rivers and hunted the vast buffalo herds. They used buffalo skins and deerskins to make *tipis* (also spelled tepees) and clothing. The Plateau tribes lived west of the prairies. They hunted, gathered food, and caught salmon and other fish in the region's rivers.

Northwest Coast peoples, such as the Tlingit and the Haida, lived on the western coast where the ocean and rivers teemed with fish and seafood. These Indians hunted game and gathered berries in the huge forests. The trees provided wood for their lodges and canoes.

The traditional lifestyles of Canada's native people began to change when Europeans arrived. Eastern groups such as the Algonquin slowly changed their way of life by trading furs for food, weapons, and traps. The various groups took different sides in the wars between the French and English, and the groups fought among themselves as well. Those who managed to survive the wars and the diseases brought by Europeans lost their land to the settlers. Some moved west, and others moved onto reservations.

The Indians on the west coast adapted to the Europeans' ways better than most. Some bought boats and started to sell their fish. Others took jobs in canneries.

The Inuit way of life began to change during the 1800's. They started working with the Europeans, first in whaling and then in fur trapping. But increased whaling, trapping, and hunting killed enormous numbers of the animals on which the Inuit depended.

By the mid-1900's, more and more Inuit were moving to communities around trading posts, government offices, and radar sites. Some Inuit found construction work, but many began to need help from the government.

## Indians and Inuit today

Today, about 780,000 Canadians are registered as Indians. Most live on reserves owned by the Indians but held in trust for them by the government. The government is responsible for education, housing, and health care on the reserves. Some Indians have protested the living conditions on the reserves. Others have sued the government to halt projects they believe would harm their rights.

About 50,000 Inuit live in Canada, mostly in Nunavut. In 1992, an agreement between Canada, the Inuit, and the Northwest Territories established a new homeland for the Inuit. *Nunavut,* meaning *our land,* became a territory in April 1999. It consists of 800,000 square miles (2 million square kilometers) carved out of the Northwest Territories.

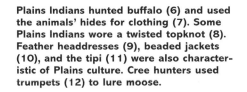

Plains Indians hunted buffalo (6) and used the animals' hides for clothing (7). Some Plains Indians wore a twisted topknot (8). Feather headdresses (9), beaded jackets (10), and the tipi (11) were also characteristic of Plains culture. Cree hunters used trumpets (12) to lure moose.

# HISTORY

About A.D. 1000, Vikings landed on the northeast coast of North America in Newfoundland. The Vikings called the region *Vinland* and founded a small settlement, but it was short-lived.

Lasting contact between Europe and Canada began in 1497, when John Cabot landed in eastern Canada and claimed the region for England. In the 1530's, Jacques Cartier sailed into the Gulf of St. Lawrence and claimed the land for France. Later, he sailed up the St. Lawrence River.

Fishing and, later, the fur trade brought the French to Canada. The growth of the colony known as New France established French culture in the country.

Meanwhile, English seamen explored Canada. In 1610, Henry Hudson sailed into the bay that was later named for him. England claimed large parts of Canada because of Cabot's and Hudson's voyages. But France also claimed this land on the basis of extensive explorations and surveys by the French explorer Samuel de Champlain. From 1689 to 1763, the British and French colonists fought each other in four wars.

The last of these wars is known as the *Seven Years' War* in Canada and Europe and as the *French and Indian War* in the United States. British troops seized Quebec City in 1759 and Montreal in 1760. The fighting ended in 1763, and France surrendered most of New France to the British.

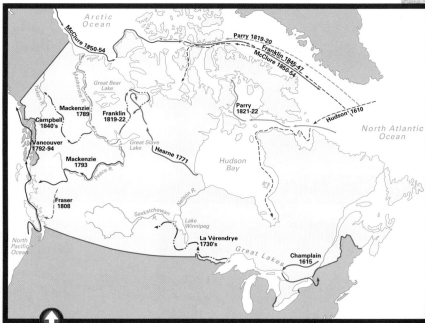

Canada was claimed by both French and English explorers. British and French colonists in Canada fought each other in four wars.

The British called their new territory *Quebec*. Prodded by Quebec Governor Guy Carleton, Britain passed the Quebec Act in 1774. This law gave the French Canadians in Quebec political and religious rights.

Other settlers in Quebec became unhappy. The British tried to solve the problem by dividing Quebec into two colonies in the Constitutional Act of 1791. Lower Canada, along the lower St. Lawrence River, remained largely French in population and law. Upper Canada, along the Great Lakes, was British in population and law.

The Canadian Guard Band plays in Quebec City's Citadel. Here in 1759, British troops under General James Wolfe captured the city from the French. Montreal fell in 1760, and New France became British.

| 1000 | Vikings reach Vinland. |
|---|---|
| 1497 | John Cabot claims Newfoundland for England. |
| 1534 | Jacques Cartier claims Canada for France. |
| 1604 | Acadia founded. |
| 1608 | Champlain founds Quebec. |
| 1642 | Montreal founded. |
| 1670 | Hudson's Bay Company founded. |
| 1689–1763 | British and French fight wars; Britain wins New France. |
| 1774 | Quebec Act gives French Canadians political and religious rights. |
| 1791 | Constitutional Act splits Quebec. |
| 1812–1815 | War with United States. |
| 1837 | Revolts break out in Upper and Lower Canada. |
| 1841 | Act of Union joins Upper and Lower Canada. |
| 1848 | Province of Canada and Nova Scotia gain self-government. |
| 1857 | Gold Rush on Fraser River. |
| 1867 | British North America Act establishes the Dominion of Canada. |
| 1869 | Riel leads Red River Rebellion. |
| 1870 | Province of Manitoba created; North West Territories (now Northwest Territories) established. |
| 1871 | British Columbia joins Confederation. |
| 1873 | Prince Edward Island joins Confederation. |
| 1885 | North West Rebellion led by Riel. |
| 1885 | Canadian Pacific Railway completed. |
| 1897–1898 | Klondike Gold Rush. |
| 1905 | Provinces of Saskatchewan and Alberta created. |
| 1914–1918 | Canadians serve in World War I. |
| 1931 | Complete independence granted. |
| 1939–1945 | Canadians serve in World War II. |
| 1949 | Newfoundland becomes 10th province. |
| 1959 | St. Lawrence Seaway opens. |
| 1962 | Trans-Canada Highway completed. |
| 1969 | Official Languages Act establishes services in both French and English. |
| 1976 | Separatists elected in Quebec. |
| 1982 | Constitution Act ends British control over amendments to Canada's Constitution. |
| 1994 | North American Free Trade Agreement takes effect. |
| 1999 | Nunavut becomes a territory. |
| 2005 | Haitian-born Michaëlle Jean is appointed governor general of Canada. |

Samuel de Champlain,
(1570?-1635)

John A. Macdonald,
(1815-1891)

Michaëlle Jean
(1957-    )

Canada began to grow as fur traders explored the northwest and thousands of immigrants arrived. Trouble brewed, however. French Canadians in Lower Canada resented the power of English-speaking Canadians. Led by Louis Papineau, many French Canadians revolted in 1837. Canadians in Upper Canada who wanted reforms followed William Mackenzie in revolt one month later. Finally, in 1841, the British united Lower and Upper Canada as the Province of Canada.

In the mid-1800's, trade and transportation routes grew as Canada prospered. But problems remained in the Province of Canada. John A. Macdonald and George Cartier campaigned for a self-governing federal union. Canadians presented a plan to the United Kingdom, and in 1867 the British North America Act established the Dominion of Canada. The new Dominion had four provinces—New Brunswick, Nova Scotia, Ontario (once Upper Canada), and Quebec (once Lower Canada). Macdonald became Canada's first prime minister. His chief goal was to extend the country to the west coast. Some settlers in a western settlement called Red River feared this expansion. They were *métis*—people of mixed European and Indian ancestry. In 1869 and again in 1885 Louis Riel led his fellow métis in unsuccessful revolts against the government.

The completion of the Canadian Pacific Railway in 1885 triggered a rush to settle the western prairies. Wilfred Laurier became the first French Canadian prime minister in 1896, and Canada prospered under his leadership. Huge wheat crops were traded in Europe, steel and textile industries and mining expanded.

With the addition of new provinces, Canada grew. In 1931, the nation won full independence. Canadians helped the Allies fight World War I (1914-1918) and World War II (1939-1945). The country's economy thrived after World War II. In 1959, the St. Lawrence Seaway opened the Great Lakes to oceangoing ships. In 1962, the Trans-Canada Highway—Canada's first ocean-to-ocean road—was completed.

In 1982, the Constitution Act ended British control over amendments to Canada's Constitution and included a bill of rights. Today, Canadian leaders continue to work for national unity, but problems between French- and English-speaking Canadians still flare.

# CONFLICT AND OPPORTUNITY

In 1967, Canada celebrated the 100th anniversary of Confederation—the union of the provinces. But maintaining that feeling of unity has often been difficult due to Canada's vast geographical area and the many differences among the people and the provinces. The proximity of the United States—with its larger population and greater economic power—keeps Canadians careful in protecting their national identity.

## Quebec and Canada

About 65,000 French colonists lived in Quebec when France lost its colony to Britain in 1763. Since that time, Canada's French population has grown to several million. Most live in Quebec, but Ontario, New Brunswick, and other regions also have many people with French backgrounds. These French Canadians, who call themselves *Québécois*, try to preserve their French language and culture.

In 1960, the Quebec government started the Quiet Revolution, a movement to defend French Canadian rights. The Quiet Revolution led some Québécois to approve of *separatism,* the demand that the province separate from Canada

and become an independent nation. The Parti Québécois organized as a separatist political party in 1968, and in 1976 it won control of Quebec's provincial government.

However, in 1980, the separatist movement suffered a major setback. Voters defeated a proposal that would have allowed Quebec to work out an agreement with the Canadian government to separate. An agreement worked out in 1987 stated that Quebec would be recognized as a distinct society in Canada. This agreement, known as the Meech Lake Accord, died in June 1990, after the provinces failed to ratify it.

In October 1992, a nationwide referendum was held on proposals to recognize Quebec as a separate society and to grant native peoples self-government. A majority of voters rejected the proposals. In September 1994, Quebecers again voted the Parti Québécois into provincial office. On October 30, 1995, Quebecers in a referendum voted narrowly against independence. In 2006, the lower house of the national Parliament passed a motion that acknowledged the unique French cultural heritage of Quebec.

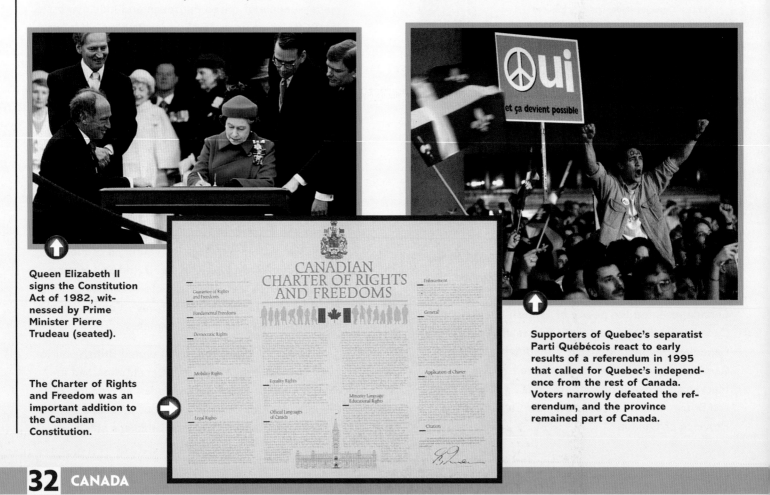

Queen Elizabeth II signs the Constitution Act of 1982, witnessed by Prime Minister Pierre Trudeau (seated).

The Charter of Rights and Freedom was an important addition to the Canadian Constitution.

Supporters of Quebec's separatist Parti Québécois react to early results of a referendum in 1995 that called for Quebec's independence from the rest of Canada. Voters narrowly defeated the referendum, and the province remained part of Canada.

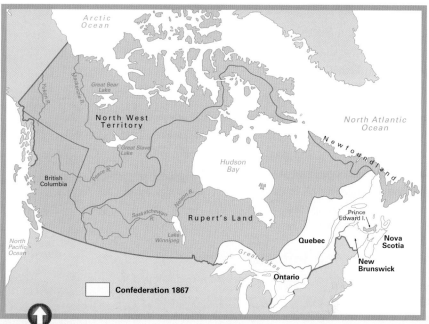

Autumn colors in Riding Mountain National Park in Manitoba display the riches of Canada's forests. They provide lumber for industry and also serve as recreation areas. Some Canadian forestland has been harmed by acid rain, which forms when water vapor in the air reacts with chemical compounds given off by cars, factories, and power plants.

Confederation 1867

In 1867, a union of British North American colonies led to the formation of the Dominion of Canada. The Dominion had four provinces—New Brunswick, Nova Scotia, Ontario, and Quebec. Britain ruled its other Canadian territories separately.

## Regionalism

Quebec is not the only Canadian province to seek more control over its own affairs. During the 1970's, the energy-rich provinces of Alberta, Saskatchewan, and British Columbia benefited from a boom in petroleum and natural gas. There was talk of independence for Alberta by westerners who objected to the national government's control over oil prices.

On the other hand, the Atlantic Provinces suffer from economic problems. This region has a lower standard of living, lower wages, and a higher rate of unemployment than any other part of Canada.

A Greenpeace activist prepares a banner aboard one of the group's vessels. Founded in Vancouver in 1971, Greenpeace uses nonviolent methods to protest seal hunts, whaling, and nuclear testing.

## Industry and the environment

Canada is rich in natural resources, including forests, petroleum, and water power. These resources have helped make Canada a leading manufacturing nation. However, recovering the resources and turning them into manufactured goods has also resulted in damage to Canada's natural environment.

For example, *acid rain,* caused by pollutants from factories as far away as the Midwestern United States, is damaging forests and streams in southeastern Canada. Oil exploration in the far northwest threatens the fragile Arctic plant and animal life. In Canada, as elsewhere in the world, there is a growing awareness of environmental problems.

# ECONOMY AND RESOURCES

Fish and, later, fur brought Europeans to Canada. Settlers also farmed and cut down trees in the forests. Some Canadians still make their living in these ways, but the major economic activities today are the service and manufacturing industries.

In August 1992, Canada, the United States, and Mexico announced a comprehensive plan for free trade across North America. Known as the North American Free Trade Agreement (NAFTA), the accord was designed to bind together the economies of the three countries in a regional trading block. After all three countries ratified the agreement, it went into effect in 1994.

## Service and manufacturing industries

About 70 percent of Canadian workers are employed in service industries. Community, social, and personal services employ more workers than any other type of Canadian industry. Such workers include teachers, doctors, data processors, and hotel employees.

Other service workers are involved in finance, insurance, or real estate. Toronto and Montreal are the leading financial centers. The federal and provincial governments

⬆

Zinc is smelted and processed in an industrial complex in Trail, British Columbia. Canada is one of the world's largest zinc-mining countries.

A truck descends to the bottom of an open pit uranium mine in McLean Lake, Saskatchewan.

participate in this area of the service economy by providing insurance that guarantees medical care for all Canadians. Service industries also include Canada's highly advanced transportation, communication, and utilities networks.

The St. Lawrence Lowlands are at the heart of Canadian manufacturing. Factories in Ontario and Quebec produce more than three-fourths of the value of all of Canada's manufactured goods.

Transportation equipment ranks first in Canada's manufacturing industry. The Canadian plants of U.S. automakers produce automobiles and trucks. Canada also manufactures aircraft and aerospace equipment.

The food-processing industry handles chiefly meat and poultry products. Other important food products include baked goods, beer, dairy products, canned fruits and vegetables, fish, and beverages.

Canadians also manufacture chemicals, medicines, machinery, metal products, steel, and paper. Mills process wood from Canada's forests into lumber, paper, plywood, and wood pulp.

Dairy farms in Quebec produce great quantities of milk on fairly small plots of land. The Prairie Provinces of Alberta, Saskatchewan, and Manitoba have vast cattle ranches and wheat farms.

A pumpjack pumps for oil in a canola field near Arcola, Saskatchewan. Petroleum production is a major economic activity in Canada.

## Mining

Canada is one of the world's top mineral exporters. Petroleum and natural gas are the country's most important minerals. The Prairie Provinces have large deposits of these valuable resources, and Alberta is Canada's leading producer. Deposits have also been found in the Arctic Islands, but these are difficult to recover and transport.

The Canadian Shield holds much of Canada's mineral wealth. Newfoundland and Labrador lead the provinces in iron ore production. Quebec also produces large amounts of iron ore, as well as gold. Ontario is Canada's leading producer of metal ores. Ontario produces much of the world's nickel, a well as some copper and gold.

Gold attracted miners to British Columbia in the 1800's and is still mined there. British Columbia is also the leading coal-mining province. Canada's other mined products include diamonds, platinum, potash, salt, sand and gravel, silver, sulfur, uranium, and zinc.

## Agriculture, fishing, and forestry

Many of the raw materials used in manufacturing come from Canada's farms, waters, and forests.

The Prairie Provinces provide most of Canada's wheat and much of its beef. Farmers in the St. Lawrence Lowlands produce a variety of goods, including grains, milk, vegetables, and fruits. Farmers in the Atlantic Provinces raise potatoes and dairy cattle.

Fishing is Canada's oldest industry. The government has had to place some restrictions on fishing in both Atlantic and Pacific waters because overfishing has threatened the population of certain fish.

Canada's forests provide logs for processing into lumber, paper, plywood, and wood pulp. Most of the forests are owned by the national and provincial governments, which lease them to private companies.

Commercial fishing boats dock at Steveston village, a historic salmon canning center in Richmond, British Columbia.

# CANADIAN CITIES

In the second half of the 1900's, Canada became an urban society. Today, about 80 percent of its people live in urban areas, and over 30 metropolitan areas have a population of more than 100,000.

## Ottawa—the capital city

Ottawa was a small lumbering town on the south bank of the Ottawa River in Ontario when Queen Victoria chose it as the capital of the United Province of Canada. With Confederation in 1867, Ottawa became the capital of the Dominion.

Today, Canada's Parliament buildings rise atop Parliament Hill. Attractive parks and scenic drives add to the beauty of the city. More than 1,000,000 people live in the Ottawa metropolitan area, and more than 100,000 of these people work for the federal government.

## Montreal—city of Mount Royal

Montreal is one of the largest French-speaking cities in the world and one of North America's most interesting cities. It lies on an island in Quebec near where the St. Lawrence and Ottawa rivers meet. A tree-covered mountain rises in its center. Montreal's downtown area includes a large network of underground stores and restaurants called the *Cite Souterrain* (Underground City). Its fascinating waterfront area has old stone buildings, cobblestone streets, and monuments.

Montreal is Canada's second largest city after Toronto. It is Canada's chief transportation hub and a major center of Canadian business, industry, culture, and education.

## Quebec—Canada's oldest city

In 1608, the French explorer Samuel de Champlain founded the settlement of Quebec where the St. Lawrence River narrows. That settlement is now a major city, the capital of Quebec Province, and an important port. Thousands visit the city each year to see "the cradle of French civilization in the New World."

Quebec has many landmarks that help give it the charm of an old European city. The Notre-Dame-des-

The Parliament buildings in Ottawa take on a golden hue as twilight settles on the city. The wooded slopes and tended lawns of Parliament Hill offer a dramatic view of Canada's capital.

Sidewalk cafes and shops attract tourists to Quebec City. Such landmarks as Place Royale, a historic square dating from the 1600's, and the Citadel, a walled fortress, are also popular with tourists.

Victoires Church, completed in 1688, stands on the site of Champlain's first settlement. The Citadel, a walled fortress, overlooks the city, and the Fairmont Château Frontenac, a castlelike hotel, rises dramatically from the skyline.

## Toronto—a dynamic city

The name *Toronto* is a Huron word meaning *meeting place*. Today, Toronto is Canada's largest metropolitan area. Lying on the northwest shore of Lake Ontario, the city is the chief manufacturing, financial, and communications center of the country, as well as one of the busiest Canadian ports on the Great Lakes.

Toronto is also noted for its cultural life. It boasts Canada's largest museum, the Royal Ontario Museum. Toronto is one of the most ethnically diverse cities in the world, with more than 150 ethnic groups that speak over 100 languages. The largest ethnic groups in the Toronto area include people of Chinese, English, Irish, and Scottish ancestry. Other large groups include Asian Indians, Italians, Pakistanis, Portuguese, and Sri Lankans. In recent years, many immigrants have come from Asia and the Caribbean Islands.

## Vancouver—gateway to the Pacific

Vancouver is the busiest port in all Canada and one of its largest cities. The people who live in the Vancouver metropolitan area make up more than half the entire population of British Columbia. The city is the province's major center of commerce, culture, industry, and transportation.

About 25 miles (40 kilometers) north of the United States, Vancouver lies in a beautiful setting near the Coast Mountains and the Pacific Ocean. The protective mountains and warm Pacific winds give Vancouver a mild climate, but the city's chief asset is its natural harbor, which never freezes. Vancouver handles nearly all of Canada's trade with Asia.

## Edmonton and Calgary—boom towns

Edmonton is the capital of the Prairie Province of Alberta and the northernmost major city in North America. Although its location makes Edmonton a distribution point for goods traveling to Alaska and the northwest, it was the discovery of oil that lured thousands of people to Edmonton. Many of Alberta's principal oil wells are within 100 miles (160 kilometers) of the city. Petroleum distributing and processing and the production of petroleum products are Edmonton's leading industries.

Calgary, in the foothills of the Rockies in southwest Alberta, grew up as a cattle town and is still a major cattle center. The Calgary Exhibition and Stampede, a world-famous rodeo, is a celebrated annual event. As in Edmonton, oil has dramatically affected the city of Calgary. The petroleum industry has made Calgary one of the country's fastest-growing cities and the oil center of Canada.

## Halifax—maritime city

The Halifax Regional Municipality in Nova Scotia is the largest municipality in Canada's Atlantic Provinces. Halifax has one of the world's largest harbors and serves as Canada's main naval base and busiest east coast port. Halifax is also one of the nation's most historic cities. The oldest parliament building in Canada stands in downtown Halifax.

Toronto is Canada's chief center of industry and finance. The CN Tower on the far left is a communications tower 1,815 feet (553 meters) high.

Vancouver was the host for the 2010 Olympic Winter Games. Skiers competed on nearby Cypress Mountain.

# THE CANADIAN WILDERNESS

Canada is a land of magnificent natural beauty. Towering mountains, crystal-clear lakes, rocky coastlines, and lush, green forests mark its wilderness areas, which cover about nine-tenths of Canada. Most of the Canadian people live near the southern border in about one-tenth of the country's land area.

To preserve some of this great wilderness and the wildlife it supports, Canada has established an extensive system of national parks. The system began in 1885, when Banff National Park was established in Alberta. Today, the system has a total of 42 national parks and park reserves, 2 national marine conservation areas, and more than 150 national historic sites. All the provinces and territories have at least one national park. Each province also has its own park system.

Canadians enjoy a wide variety of recreational activities, such as canoeing, fishing, hiking, and snowshoeing. Many of these activities take place in their national parks.

## Eastern wilderness

The Atlantic, or Maritime, Provinces include a number of superb natural areas. Gros Morne National Park, for example, on the west coast of Newfoundland, has fiordlike lakes, waterfalls, a rugged seacoast, and the scenic Long Range Mountains. At Sir Richard Squires Memorial Park, also in Newfoundland, Atlantic salmon make spectacular jumps up the Big Falls to reach their egg-laying grounds upstream. In Cape Breton Highlands National Park in Nova Scotia, the Cabot Trail highway hugs rugged coastline and dips into deep canyons.

A well-known wilderness area in Quebec is the Gaspé Peninsula. This rugged arm of land attracts artists, hikers, and climbers. Bonaventure Island, off the coast of Gaspé, is one of the largest water-bird refuges open to visitors. The Laurentian Mountains, or Laurentides, form the southeastern edge of the Canadian Shield in Quebec. Low ranges in the provincial parks of Laurentides and Mont Tremblant attract skiers.

**National parks of Canada**

Quttinirpaaq

Aulavik
Ivvavik
Vuntut
Sirmilik
Auyuittuq
Tuktut Nogait
Kluane
Nahanni
Ukkusiksalik
Torngat Mountains
Wood Buffalo
Wapusk
Terra Nova
Gros Morne
Gwaii Haanas
Elk Island
Prince Albert
Mingan Archipelago
Glacier
Jasper
Forillon
Cape Breton Highlands
Mt. Revelstroke
Banff
Kouchibouguac
Pacific Rim
Yoho
Kootenay
Riding Mountain
Prince Edward Is.
Fundy
Gulf Islands
Waterton Lakes
La Mauricie
Kejimkujik
Grasslands
Pukaskwa
St. Lawrence Islands
Bruce Peninsula
Georgian Bay Islands
Point Pelee

Ontario has more people than any other Canadian province, but almost all of them live in the extreme southeast section. Huge forests cover much of the land, and wild flowers are plentiful. Woodland caribou, moose, shy white-tailed deer, and black bears roam these northern woodlands. Ontario has 5 national parks, a national marine park in Georgian Bay, and more than 250 provincial parks. The densely forested Quetico Provincial Park, for example, draws many canoeists to its beautiful lakes.

## Western wilderness

Stretching west, Canada's landscape changes, and the wilderness areas and parks change too. In Saskatchewan, Grasslands National Park protects the short-grass prairie that is native to the area, as well as the wildlife it supports. Riding Mountain National Park in Manitoba features grasslands, lakes, and forests around a ridge that rises above the flat, rolling land.

Farther west, the Canadian Rockies erupt boldly from the prairie. Some of the best-known parks in the world are located there. Banff National Park and Jasper National Park attract millions of tourists to their peaks, glaciers, hot springs, and resorts.

Alberta also lies along three of the major North American flyways used by birds migrating between their winter and summer homes. Wetlands are managed to help ensure the success of nesting waterfowl. Wood Buffalo National Park in the northeastern part of the province has one of the world's largest herds of free-roaming wood bison and the nesting grounds of rare whooping cranes.

The wilderness of British Columbia is also a favorite with tourists from all across Canada and the United States. Gwaii Haanas National Park on the Queen Charlotte Islands protects rare plant and animal life. Pacific Rim National Park on Vancouver Island features sea lions. British Columbia also has national and provincial parks throughout its mountain areas.

## The Far North

The Far North of Canada is itself one vast wilderness. The territories make up more than one-third of Canada's land area, yet less than 1 percent of the country's people live there. Forest-covered mountains spread over Yukon and part of the Northwest Territories. Kluane National Park in Yukon contains the highest peak in Canada—Mount Logan. Large glaciers lie in its mountains, and Dall sheep roam their slopes. Nahanni National Park in the Northwest Territories features deep canyons, hot springs, and the Virginia Falls.

Much of the rest of the Far North is *tundra*—cold, dry, treeless land that is home to caribou, bears, foxes, and other northern animals. Although this environment is harsh, it is also fragile.

**The forests and flat tundra of Canada's wilderness areas provide a habitat for large mammals such as moose (1), caribou (2), and bears (3). Small fur-bearing mammals include raccoons (4), gray squirrels (5), and beavers (6). Birds include cardinals (7), snowy owls (8), Canada geese (9), hairy woodpeckers (10), and mallard ducks (11). The many lakes and rivers in the country are home to northern pike (12), perch (13), and rainbow trout (14).**

# CAPE VERDE

The republic of Cape Verde *(kayp VURD)* is an island nation off the western coast of Africa. Its 15 islands were formed by volcanic eruptions that took place 2-1/2 million to 65 million years ago. Most of the islands are ruggedly mountainous, with steep cliffs rimming their coastlines.

Portuguese explorers discovered the uninhabited islands of Cape Verde about 1460. Settlers came about two years later, planting cotton, fruit trees, and sugar cane. They also brought slaves from the African mainland to work the land.

The slave trade eventually became the most important economic activity on the islands, and Cape Verde prospered. Slaves were "trained" on plantations there before being shipped elsewhere. But the slave trade declined in the late 1600's, and prosperity disappeared with it.

Portugal ruled Cape Verde and what is now Guinea-Bissau on mainland Africa under one government from 1579 until 1879, when each became a separate Portuguese province. The African Party for the Independence of Guinea and Cape Verde, or PAIGC (its initials in Portuguese), formed in 1956 to fight for full independence, which Cape Verde achieved in 1975.

Until 1991, the PAIGC—now the African Party for the Independence of Cape Verde (PAICV)—was the only legal political party in the country. In that year, the PAICV was voted out of office in multiparty elections. But it won the elections in 2001, 2006, and 2011.

Cape Verde, an underdeveloped nation, relied almost entirely on Portugal for economic support before its independence. Since 1975, Cape Verde has received food aid from the United Nations and financial aid from various countries. Service industries—especially commerce, tourism, transportation, and communication—have become the most important economic activity. Cape Verde's cool breezes, white sand beaches, and clean, sparkling water offer a pleasant destination for vacationers.

## FACTS

| | |
|---|---|
| Official name: | Republica de Cabo Verde (Republic of Cape Verde) |
| Capital: | Praia |
| Terrain: | Steep, rugged, rocky, volcanic |
| Area: | 1,557 mi² (4,033 km²) |
| Climate: | Temperate; warm, dry summer; precipitation meager and very erratic |
| Main rivers: | N/A |
| Highest elevation: | Pico, a volcano on Fogo Island, 9,281 ft (2,829 m) |
| Lowest elevation: | Atlantic Ocean, sea level |
| Form of government: | Republic |
| Head of state: | President |
| Head of government: | Prime minister |
| Administrative areas: | 17 concelhos (municipalities) |
| Legislature: | Assembleia Nacional (National Assembly) with 72 members serving five-year terms |
| Court system: | Supremo Tribunal de Justia (Supreme Tribunal of Justice) |
| Armed forces: | 1,200 troops |
| National holiday: | Independence Day - July 5 (1975) |
| Estimated 2010 population: | 566,000 |
| Population density: | 364 persons per mi² (140 per km²) |
| Population distribution: | 60% urban, 40% rural |
| Life expectancy in years: | Male, 68; female, 74 |
| Doctors per 1,000 people: | 0.5 |
| Birth rate per 1,000: | 27 |
| Death rate per 1,000: | 6 |
| Infant mortality: | 35 deaths per 1,000 live births |
| Age structure: | 0-14: 37%; 15-64: 57%; 65 and over: 8% |
| Internet users per 100 people: | 21 |
| Internet code: | .cv |
| Languages spoken: | Portuguese (official), Crioulo (a blend of Portuguese and West African words) |
| Religions: | Roman Catholic mixed with indigenous beliefs, Protestant |
| Currency: | Cape Verdean escudo |
| Gross domestic product (GDP) in 2008: | $1.77 billion U.S. |
| Real annual growth rate (2008): | 6.0% |
| GDP per capita (2008): | $3,377 U.S. |
| Goods exported: | Clothing, fish, footwear, fuels |
| Goods imported: | Construction materials, food, fuels, machinery, transportation equipment |
| Trading partners: | Brazil, Italy, Netherlands, Portugal, Spain |

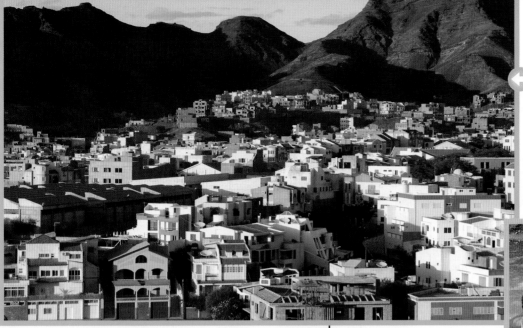

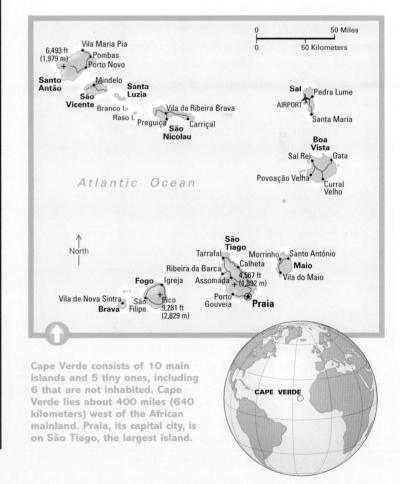

Cape Verde children enjoy a swim in the Atlantic waters that surround the islands.

Cape Verde has a warm, dry climate, and the lack of rainfall makes most of the land too dry to farm. Nevertheless, agriculture employs about one-fourth of the country's workers. The chief crops include sugar cane; bananas and other fruits; and such vegetables as beans, corn, sweet potatoes, and tomatoes. Poor rainfall and occasional droughts make farm production unpredictable and force Cape Verde to import most of its food.

Fishing and mining play a small role in Cape Verde's economy. Tuna is the main catch. The country's mining industry produces salt and *pozzuolana,* a volcanic rock used by the cement industry. Both of these products are exported.

Because of their country's underdeveloped economy, the people of Cape Verde have a low standard of living, and many are undernourished due to drought and low agricultural production. Since the mid-1900's, hundreds of thousands of Cape Verdeans have emigrated to Brazil, Portugal, the United States, and other countries.

Today, the population of the islands is about 566,000. About 70 percent of the people have mixed African and Portuguese ancestry, and most of the rest are descended from Africans. Cape Verdeans speak a local dialect called Crioulo, which is based on Portuguese and African languages. Most are Roman Catholics, but many practice *animism*—the belief that everything in nature has a soul.

Cape Verde consists of 10 main islands and 5 tiny ones, including 6 that are not inhabited. Cape Verde lies about 400 miles (640 kilometers) west of the African mainland. Praia, its capital city, is on São Tiago, the largest island.

# CENTRAL AFRICAN REPUBLIC

The Central African Republic is a land-locked country in the center of the African continent. Most of the country is a vast, rolling plateau, covered by grass and scattered trees and broken by river valleys. The northeast is arid, but thick rain forests blanket the southwest.

Rivers are the most important transportation routes in the Central African Republic. The nation has no railroads, and many roads are impassable during the rainy season.

## Economy

The Central African Republic is a poor, underdeveloped country, where most of the people farm for a living. Some people also hunt and fish to feed their families. Some farmers raise livestock in regions where there are no *tsetse flies*—insects that spread a disease called *sleeping sickness*. Some plantations raise coffee, cotton, and rubber for export.

The country has a few factories, including a textile mill. Diamond mining is the only important mining industry.

Most of the people of the Central African Republic belong to one of the many African ethnic groups in the country. French is the official language, but Sango, one of the many African languages used in the country, is the most widely spoken. Most adults are unable to read or write.

## History

The people in what is now the Central African Republic lived in small settlements before the arrival of Europeans in the 1800's. The Europeans brought slave raids and turmoil to the region. In 1889, France set up an outpost at Bangui, which is now the capital. In 1894, the French made the region a territory called Ubangi-Shari, which later became part of French Equatorial Africa.

## FACTS

| | |
|---|---|
| Official name: | Republique Centrafricaine (Central African Republic) |
| Capital: | Bangui |
| Terrain: | Vast, flat to rolling, monotonous plateau; scattered hills in northeast and southwest |
| Area: | 240,535 mi² (622,984 km²) |
| Climate: | Tropical; hot, dry winters; mild to hot, wet summers |
| Main rivers: | Ubangi, Mbomou, Sangha, Kotto |
| Highest elevation: | Mont Ngaoui, 4,658 ft (1,420 m) |
| Lowest elevation: | Ubangi River, 1,099 ft (335 m) |
| Form of government: | Republic |
| Head of state: | President |
| Head of government: | Prime minister |
| Administrative areas: | 14 prefectures, 2 prefectures economiques (economic prefectures), 1 commune |
| Legislature: | Assemblee Nationale (National Assembly) with 105 members serving five-year terms |
| Court system: | Cour Supreme (Supreme Court), Constitutional Court |
| Armed forces: | 3,200 troops |
| National holiday: | Republic Day - December 1 (1958) |
| Estimated 2010 population: | 4,574,000 |
| Population density: | 19 persons per mi² (7 per km²) |
| Population distribution: | 62% rural, 38% urban |
| Life expectancy in years: | Male, 44; female, 44 |
| Doctors per 1,000 people: | Less than 0.05 |
| Birth rate per 1,000: | 36 |
| Death rate per 1,000: | 18 |
| Infant mortality: | 102 deaths per 1,000 live births |
| Age structure: | 0-14: 42%; 15-64: 54%; 65 and over: 4% |
| Internet users per 100 people: | 0.4 |
| Internet code: | .cf |
| Languages spoken: | French (official), Sangho (lingua franca and national language), tribal languages |
| Religions: | Indigenous beliefs 35%, Protestant 25%, Roman Catholic 25%, Muslim 15% |
| Currency: | Coopération Financière en Afrique Centrale franc |
| Gross domestic product (GDP) in 2008: | $2.02 billion U.S. |
| Real annual growth rate (2008): | 3.5% |
| GDP per capita (2008): | $485 U.S. |
| Goods exported: | Coffee, cotton, diamonds, wood products |
| Goods imported: | Food, machinery, petroleum products, pharmaceuticals, vehicles |
| Trading partners: | Belgium, France, Indonesia, Italy, Netherlands |

Pygmy people of the rain forest live by hunting wild animals and gathering plants.

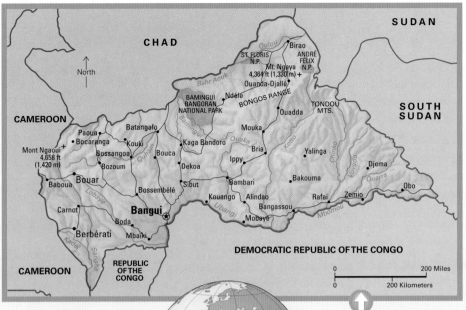

The Central African Republic consists largely of a vast plateau broken by deep river valleys.

In 1958, Ubangi-Shari gained internal self-government as the Central African Republic, and the country became fully independent on Aug. 13, 1960. David Dacko became the new nation's first president. In 1966, Dacko was overthrown by army officers, and Jean-Bedel Bokassa, head of the army, took over. In 1972, Bokassa was named president for life, and in 1976 he declared himself emperor. He changed the name of the country to the Central African Empire.

When Bokassa was overthrown by supporters of Dacko in 1979, the nation's name was changed back to the Central African Republic. In March 1981, Dacko was elected president. But in September, he was overthrown by Chief General André Kolingba. In 1993, the constitution was revised to permit a multiparty democracy. In presidential and parliamentary elections held that year, Ange-Félix Patasse became president.

During the mid-1990's, rebel soldiers staged several revolts against the government of Patasse. Hundreds of people were killed and forced from their homes during the fighting. A cease-fire agreement ended the hostilities in mid-1997.

In March 2003, rebels led by General François Bozizé seized control of the government. Bozizé suspended the country's constitution, dissolved the parliament, and declared himself president. In 2004, voters approved a new constitution. In 2005, under the new constitution, Bozizé was elected president. He was reelected in 2011.

The Ubangi River forms much of the Central African Republic's southern boundary with the Democratic Republic of the Congo.

# CHAD

Chad is a large and thinly populated country in north-central Africa. It is a landlocked nation, but Lake Chad lies on its western border. Today, Chad is one of the poorest nations in the world. Desert covers much of Chad in the north. A small fertile region in the south was called "Useful Chad" by the French because the rest of the country had infertile soil and few other resources.

Between the northern desert and the southern fertile strip lies a *savanna,* or grassy plain. The savanna is part of a larger region called the Sahel. Severe droughts hit the Sahel from the late 1960's to the 1980's, destroying crops and killing livestock. Famine followed, and millions of Africans died.

Since 2000, Chad has begun to develop its oil resources, and oil exports began in 2004. The country hopes to use oil revenue to pay for social development programs.

## Modern life

Sharp religious, social, and economic differences between the people of the north and south have kept Chad in an almost constant state of civil war since the mid-1960's.

Most of the people in northern Chad are Muslims, while most of the people in the south follow African or Christian religions. The Sara are the largest ethnic group in the south. Most of the country's schools and limited industry are in the south.

The huge gap in education and economic development, along with the religious difference, increases the tension between the north and south. People in the north believe they do not have equal opportunity. Meanwhile, southern people recall that, for hundreds of years, northern raiders seized the southern Chadians as slaves.

## History

A kingdom called Kanem arose along the desert trade routes northeast of Lake Chad about the A.D. 700's. Islam was introduced into the region around

## FACTS

| | |
|---|---|
| Official name: | Republique du Tchad (Republic of Chad) |
| Capital: | N'Djamena |
| Terrain: | Broad, arid plains in center, desert in north, mountains in northwest, lowlands in south |
| Area: | 495,755 mi² (1,284,000 km²) |
| Climate: | Tropical in south, desert in north |
| Main rivers: | Chari, Logone, Bahr Salamat |
| Highest elevation: | Emi Koussi, 11,204 ft (3,415 m) |
| Lowest elevation: | Lake Chad, 922 ft (281 m) |
| Form of government: | Republic |
| Head of state: | President |
| Head of government: | Prime minister |
| Administrative areas: | 22 regions |
| Legislature: | National Assembly with 155 members serving four-year terms |
| Court system: | Supreme Court, Court of Appeal, Criminal Courts, Magistrate Courts |
| Armed forces: | 26,000 troops |
| National holiday: | Independence Day - August 11 (1960) |
| Estimated 2010 population: | 11,678,000 |
| Population density: | 24 persons per mi² (9 per km²) |
| Population distribution: | 73% rural, 27% urban |
| Life expectancy in years: | Male, 46; female, 48 |
| Doctors per 1,000 people: | Less than 0.05 |
| Birth rate per 1,000: | 44 |
| Death rate per 1,000: | 16 |
| Infant mortality: | 106 deaths per 1,000 live births |
| Age structure: | 0-14: 46%; 15-64: 51%; 65 and over: 3% |
| Internet users per 100 people: | 1.2 |
| Internet code: | .td |
| Languages spoken: | French (official), Arabic (official), Sara (in south), more than 120 different languages and dialects |
| Religions: | Muslim 51%, Christian 35%, animist 7%, other indigenous beliefs 7% |
| Currency: | Coopération Financière en Afrique Centrale franc |
| Gross domestic product (GDP) in 2008: | $8.39 billion U.S. |
| Real annual growth rate (2008): | 1.7% |
| GDP per capita (2008): | $792 U.S. |
| Goods exported: | Cotton, gum arabic, livestock, oil |
| Goods imported: | Food, industrial goods, machinery, textiles, transportation equipment |
| Trading partners: | Cameroon, France, United States |

1100. Later, two smaller kingdoms—Baguirmi and Ouaddai—developed near Kanem in the 1500's and 1600's. All three kingdoms became powerful and prosperous by trading goods and slaves they had captured from the Sara.

In the late 1800's, France claimed Chad as its own. In 1920, Chad became a French colony. The Sara people suffered more than any other group under colonial rule. When Chad gained its independence from France in 1960, a Sara government took over.

In 1962, a group of northern rebels formed a group called the *Front de Libération National* (National Liberation Front), or *Frolinat*. When civil war broke out between Frolinat and government troops, Chad's government turned to France for help. Frolinat received aid from Libya. A military coup took over the government in 1975. Fighting continued into the 1980's. The groups finally agreed to a truce in 1987. However, some Libyan troops remained in Chad until 1994, when the United Nations settled a final dispute.

In 1990, a rebel group overthrew Hissene Habré's government. Chad set up an interim government in 1993. In 1996, the country adopted a new constitution and held multiparty presidential elections, which were followed by legislative elections in 1997.

In 1998, the Movement for Democracy and Justice in Chad, based in northern Chad, began an armed rebellion against the government. The government and the rebels signed several peace agreements, but violence continued to flare until late 2003. Rebel uprisings reignited in 2006, 2007, and 2008.

**Chad, a large landlocked African country, takes its name from the great lake that lies on its western border. Differences between the peoples of the north and the south led to civil war in Chad in 1998. Because of the conflict and the nation's lack of resources, Chad is extremely poor.**

Dromedary camels gather at a *guelta* (desert watering hole) on the Ennedi Plateau in northeastern Chad.

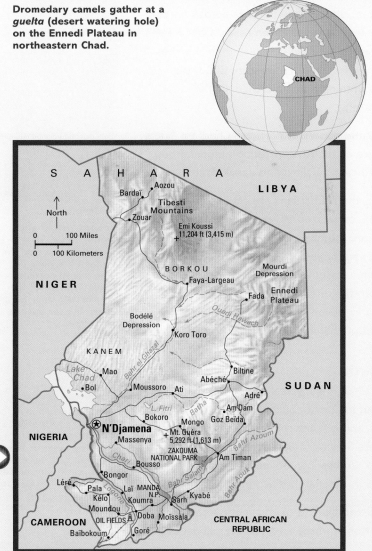

# LAND AND PEOPLE

In Chad, differences between the north and the south exist not only in the landscape, but also in the people. However, the north and south alike are affected by Chad's under-developed economy.

Only about a fourth of all Chadian adults can read and write. The country has no railroads and few paved roads. There is limited phone and postal service. The chief manufactured products include beer, cigarettes, construction material, cotton products, meatpacking products, and soap. But civil war has closed down most of these businesses. The people of Chad rely heavily on aid from France and other countries.

## The north
In the great Sahara that stretches across northern Chad, temperatures often reach 120° F (49° C), and the annual rainfall is less than 5 inches (13 centimeters). In northwestern Chad, the Tibesti Mountains rise more than 11,000 feet (3,000 meters).

Most of the people in northern Chad are Arabs or members of the African Toubou ethnic group. The majority are Muslims and speak Arabic.

Most northern Chadians are cattle traders who raise cattle, camels, goats, and sheep. While milk and meat are the basis of their diet, they also eat dates and vegetables grown in oases. The northern Chadians travel through the desert in small groups with their herds. They construct tents out of sticks and woven mats. In the north, less than 10 percent of the children go to school.

## The south
Southern Chad is a tropical forest, warm and much wetter than the north. Such wild animals as lions, elephants, and giraffes roam parts of the south. Large rivers flow northeast across the region into Lake Chad. Cranes, crocodiles, and hippopotamuses live in the marshes around the lake.

The soil and climate of the south are ideal for raising cotton, and the area has the richest farm-

Toubou gather palm fronds to weave into baskets and mats. The Toubou are nomads who live in northern Chad. They roam the desert with their herds, searching for pasture and water.

The town of Faya-Largeau south of the Tibesti Mountains is a crossing point of two Saharan caravan routes. Here, northern nomads trade their livestock for food. Their robes and turbans are traditional clothing in northern Chad.

Round granaries (grain storehouses) are a common sight in southern Chad. In this small but fertile region, the soil and climate are good for growing cotton and other crops. Southern Chad may further benefit from future development of oil deposits in the region.

Chadians fish in Lake Chad as well as southern rivers, but fishing is seasonal. Deposits of natron, a mineral, lie near the lake, and uranium is found in the north. These resources may one day improve Chad's economy.

land in Chad. Most Chadians live in this fertile region.

Most of the people of southern Chad are members of African ethnic groups, including the Sara—the largest group. The majority of southern Chadians follow traditional African religions, but many have converted to Christianity. Sara is the most widely spoken language in the south, but many other languages are spoken too.

Most southern Chadians farm for a living. Their main cash crops are cotton and peanuts. Farmers raise millet, sorghum, and rice for their own use. Meals occasionally include vegetables, fish, or meat.

In the rural south, Chadians live in round huts made of adobe brick or dried mud and covered with a straw roof. Some huts are made entirely of straw.

About a fourth of Chad's people live in N'Djamena or other cities, located mainly in the south. Most of the country's business people, teachers, traders, and government workers live in the south.

In the early 2000's, the World Bank and the U.S. oil company Exxon Mobil Corporation funded construction of a pipeline from southern Chad through Cameroon to the Gulf of Guinea. In return, Chad agreed to use most of the oil revenues for social development projects. Oil production came on-stream in 2003.

The most striking geographical feature of Chile is its unusual shape—it is 10 times as long as it is wide. Like a ribbon stretching down the western coast of South America, Chile is about 2,650 miles (4,265 kilometers) long but only 265 miles (427 kilometers) wide at its widest point. But to those who have explored Chile, its shape is only part of its fascination.

## Spectacular scenery

Because of its great length, Chile is a land of extremes in both landscape and climate. The Atacama Desert in the north is one of the driest places on the earth. Here, in this re-mote, barren land—free from the interference of bright artificial lights and polluted air—as-tronomers at the Cerro Tololo Inter-American Observatory study the stars and planets.

Far from the northern desert, the lonely, windswept coast of the icy southern regions is an equally desolate but different landscape of snow-capped volcanoes, thick forests, and glaciers. To the east, the mighty Andes Moun-tains extend along Chile's border with Ar-gentina and Bolivia. Nestled between the Andes and the lower mountains on the west coast lies the Central Valley. Several rivers flow through this mild, fertile region—the heartland of Chile and the center of its pop-ulation.

## A history of conflict

Isolated from the rest of the world by moun-tains, deserts, and glaciers, the Chilean peo-ple have a long history of resisting invasion. The early Indian inhabitants of the area fought bravely to protect their land—first against Inca forces and later against the Spanish conquistadors.

Long before the first Europeans arrived, several different Indian groups inhabited what is now Chile. The Atacama, Diaguita, and other tribes lived along the north coast and at the southern edge of the Atacama Desert. Chile's largest Indian group was the Mapuche, a fierce, warlike tribe that inhabited the Central Valley.

# CHILE

In the late 1400's, much of northern Chile fell under the rule of the Inca empire. However, their triumphant march south to the Central Valley ended in defeat when the Mapuche routed the Inca armies.

After the Spanish conquistadors defeated the Inca and seized their gold and silver, they headed south in search of more riches. In 1535, Diego de Almagro and his men reached the area around present-day Santiago. However, they turned back after finding only scattered Indian settlements with no gold and silver.

In 1540, another Spaniard, Pedro de Valdivia, led an expedition from what is now Peru to Chile's Central Valley. On Feb. 12, 1541, Valdivia and his men founded Santiago. The Mapuche, whom the Spaniards called *Araucanians*, destroyed the settlement only six months later. The Spaniards rebuilt Santiago and later founded the cities of La Serena, Valparaíso, Concepción, Valdivia, and Villarrica.

In the end, however, the Spaniards fared no better than the Inca against the Mapuche. In 1553, the Mapuche killed Valdivia and most of his men in battle. These fierce warriors never weakened in their resistance to Spanish rule, making Chile a bloody battleground for almost 300 years.

Spain ruled Chile from the 1500's to the early 1800's. Chile was part of a large Spanish colony called the Viceroyalty of Peru, which included other parts of Spanish South America. During the colonial period, the Roman Catholic Church sent missionaries to Chile to convert the Indians to Christianity. In time, the church became a powerful institution in the colony.

Early in the 1800's, Chile joined the other New World colonies in rebellion against Spain. After an eight-year struggle, led first by José Miguel Carrera and later by Bernardo O'Higgins and the Argentine general José de San Martín, Chile won its independence in 1818. Chile's 1833 constitution remained in effect until 1925 and gave the country a strong central government.

# CHILE TODAY

National elections in 1989 brought democratic government back to Chile after 16 years of harsh military rule. Patricio Aylwin Azócar, the candidate who represented a coalition of 17 political groups, won the people's vote. Aylwin's victory reawakened the hope of democracy for many Chileans, but serious economic and social problems have remained an obstacle to the nation's progress. Chile's lower-class people are still in desperate need of jobs, decent housing, health care, and better nutrition.

## Allende's Chile

During the 1960's, President Eduardo Frei Montalva gave Chile's people a glimmer of hope for a better life. After he was elected in 1964, Frei introduced limited land reforms and improved standards of health and housing. He also convinced the American owners of Chilean copper mines to give his government greater control of the nation's copper industry.

Leftist groups thought Frei's measures were not strong enough, while conservatives condemned them as being too radical. In 1970, Salvador Allende Gossens, who headed a coalition of Communist and other left-wing parties, was elected president. Allende became the first Marxist to win a democratic presidential election in the Western Hemisphere.

Allende attempted to reshape the nation along socialist lines. He increased the minimum wage and redistributed land among rural farmers. The Allende government also took over many of Chile's private banks, the huge copper mines, and numerous industries.

But when the Allende government increased the minimum wage, it was also trying to hold down the prices of consumer goods. As a result, food shortages became widespread, and inflation soared from about 20 percent in 1971 to more than 350 percent in 1973. Strikes became widespread, productivity declined, and many educated professional people fled the country.

## FACTS

| | |
|---|---|
| Official name: | Republica de Chile (Republic of Chile) |
| Capital: | Santiago |
| Terrain: | Low coastal mountains; fertile central valley; rugged Andes in east |
| Area: | 291,930 mi² (756,096 km²) |
| Climate: | Temperate; desert in north; Mediterranean in central region; cool and damp in south |
| Main rivers: | Loa, Maipo, Maule, Bío-Bío |
| Highest elevation: | Ojos del Salado, 22,572 ft (6,880 m) |
| Lowest elevation: | Pacific Ocean, sea level |
| Form of government: | Republic |
| Head of state: | President |
| Head of government: | President |
| Administrative areas: | 15 regiones (regions) |
| Legislature: | Congreso Nacional (National Congress) consisting of the Senado (Senate) with 38 members serving eight-year terms and the Camara de Diputados (Chamber of Deputies) with 120 members serving four-year terms |
| Court system: | Corte Suprema (Supreme Court), Constitutional Tribunal |
| Armed forces: | 60,600 troops |
| National holiday: | Independence Day - September 18 (1810) |
| Estimated 2010 population: | 17,088,000 |
| Population density: | 59 persons per mi² (23 per km²) |
| Population distribution: | 88% urban, 12% rural |
| Life expectancy in years: | Male, 75; female, 81 |
| Doctors per 1,000 people: | 1.1 |
| Birth rate per 1,000: | 15 |
| Death rate per 1,000: | 5 |
| Infant mortality: | 8 deaths per 1,000 live births |
| Age structure: | 0-14: 24%; 15-64: 67%; 65 and over: 9% |
| Internet users per 100 people: | 33 |
| Internet code: | .ci |
| Languages spoken: | Spanish (official), Mapudungun, German, English |
| Religions: | Roman Catholic 70%, Evangelical 15.1%, Jehovah's Witness 1.1%, other 13.8% |
| Currency: | Chilean peso |
| Gross domestic product (GDP) in 2008: | $169.57 billion U.S. |
| Real annual growth rate (2008): | 4.0% |
| GDP per capita (2008): | $10,116 U.S. |
| Goods exported: | Beverages, chemical products, copper and other minerals, fish products, fruits, wood products |
| Goods imported: | Chemicals, electronic equipment, machinery, motor vehicles, petroleum |
| Trading partners: | Argentina, Brazil, China, Japan, United States |

Michelle Bachelet waves to the crowd as she arrives at the presidential palace after being sworn in as Chile's first woman president on March 11, 2006.

## Military government

In 1973, military leaders overthrew Allende's government. After the coup, they claimed that Allende committed suicide rather than resign. The military leaders formed a junta, led by Augusto Pinochet Ugarte, to rule Chile. The Pinochet government imprisoned many of its opponents, dissolved the Congress, restricted freedom of the press, and banned political parties. It returned many of the state-owned industries to private control and checked inflation. In 1988, the people rejected Pinochet's bid to extend his term in office, paving the way for a return to democratic government in 1989. Chile's economy boomed during the 1990's.

In January 2000, Chile's voters elected Ricardo Lagos Escobar of the Socialist Party as president. During the 2000's, courts brought charges against Pinochet for covering up kidnappings and murders committed shortly after he gained power in 1973. Pinochet died in 2006 before going to trial.

In 2006, Chileans elected Michelle Bachelet of the Socialist Party as the country's first woman president. Twenty years of leftist government ended in 2010, when voters elected conservative candidate and businessman Sebastián Piñera.

Chile extends along the west coast of South America from the Peruvian border to Cape Horn. Easter Island and the Juan Fernández Islands are also part of Chile. The country claims a portion of the Antarctic Peninsula, although few other countries acknowledge the claim.

PERU

BOLIVIA

Putre

Arica
LAUCA N.P.
VOLCAN ISLUGA N.P.

Pisagua

Iquique

Tocopilla
María Elena
Pedro de Valdivia
Chuquicamata
Calama
Salar de Atacama

**Antofagasta**

Taltal

*South Pacific Ocean*

PAN DE AZÚCAR N.P.
Pueblo Hundido
Chañaral
Potrerillos

*Copiapó*
Copiapó
Ojos del Salado 22,572 ft (6,880 m)

Freirina
Vallenar
Norte Chico

La Serena
Coquimbo
Point Lengua de Vaca
Andacollo
Ovalle
BOSQUE FRAY JORGE N.P.
Illapel

STATUE OF CHRIST OF THE ANDES
*Aconcagua*
San Felipe
Uspallata Pass

**Viña del Mar**
**Valparaíso**
*Mapocho*
San Antonio
**Santiago**
Mt. Tupungato 22,310 ft (6,800 m)

**Rancagua**
EL TENIENTE COPPER MINE
San Fernando
*Maipo*

North

Curicó
**Talca**
Linares

**Talcahuano**
Point Lavapié
**Concepción**
Chillán
Los Ángeles

ARGENTINA

**Temuco**
*Tolhén*

Valdivia
Point Galera
La Unión

Osorno

*Lake Llanquihue*
Mt. Tronador 11,660 ft (3,554 m)

**Puerto Montt**
Ancud
Chiloé I.
Puerto Quellón

*Gulf of Corcovado*
QUEULAT N.P.

ISLA MAGDALENA N.P.
Los Chonos Archipelago
Puerto Aisén
Coihaique
Puerto Chacabuco

Taitao Peninsula
Mt. San Valentín 13,314 ft (4,058 m)
LAGUNA SAN RAFAEL N.P.

*South Pacific Ocean*

*Penas Gulf*

Wellington I.
BERNARDO O'HIGGINS N.P.

*South Atlantic Ocean*

TORRES DEL PAINE N.P.
Puerto Natales

Reina Adelaida Archipelago
Desolación I.
Riesco I.
Santa Inés I.
HERNANDO DE MAGALLANES N.P.
Punta Arenas
Porvenir
*Strait of Magellan*
Island of Tierra del Fuego

ALBERTO DE AGOSTINI N.P.
Puerto Williams
CAPE HORN N.P.
Cape Horn

CHILE

0   100   200 Miles
0   100   200   300 Kilometers

# ENVIRONMENT

A small farming settlement in Chile's Lake Country stands in the shadow of the snow-capped Andes. The Lake Country's mountainous landscape, dense forests, and crystal-clear lakes have led tourists to call it Little Switzerland.

Chile is a land of extreme contrasts in its climate, geography, and plant life. Its mainland landscape ranges from barren deserts to green valleys and lush rain forests, from treeless plains to gleaming glaciers. Far off Chile's western coast in the South Pacific Ocean lies Easter Island, its three extinct volcanoes a silent reminder of the island's fiery origins.

Chile's three main land regions—the Northern Desert, the Central Valley, and the Archipelago—have their own unique characteristics. However, they all share one common feature—the frequent occurrence of violent natural activity. Flash floods caused by the sudden melting of mountain snows often endanger lives and property in the farming villages of the Andean valleys. On the Pacific coast, storms and unpredictable offshore currents are a constant challenge to sailors and fishing crews.

But the country's most serious natural activities are earthquakes. Chile lies along a major earthquake belt, and well over 100 quakes, many accompanied by massive tidal waves, have been recorded since 1575. In the early 1900's, an earthquake almost leveled the city of Valparaíso, and 20,000 people were killed by a quake that struck Concepción in 1939. In 2010, a powerful quake struck central Chile, killing more than 500 people and injuring some 12,000 others.

## Deserts and valleys

Chile's Northern Desert region stretches from the Peruvian border in the north to the Aconcagua River, just north of Valparaíso. It includes the Atacama Desert, a region so arid that a 1971 rainfall was the first precipitation recorded in the desert town of Calama in 400 years.

Because of the lack of rainfall, the Atacama Desert has almost no plant life, except for the typical tola desert brush. Any moisture comes from subsurface water. The desert oases support small farming communities.

In the south, the Atacama gradually gives way to a slightly less arid area called the *Norte Chico* (Little North). Farmers there raise livestock and grow crops in irrigated river valleys.

In contrast to the desolate Northern Desert, the Central Valley is Chile's chief population, agricultural, and industrial center. Orchards, vineyards, pastures, and croplands cover much of the Central Valley.

A few charred tree limbs are all that remain of a forest that stood in the path of a lava flow from Llaima Volcano. Located in the Central Valley, Llaima is one of several volcanoes in the western Andes whose eruptions have frequently altered Chile's landscape.

Icebergs and ice floes form when massive glaciers plunge into the sea, creating a floating hazard for ships along Tierra del Fuego's coastline.

The region south of the Bío-Bío River, known as the Lake Country, is especially beautiful. Snow-capped volcanoes rise on the western Andean slopes, while thick forests of laurel and magnolia trees blanket the mountains. Sparkling lakes and deep valleys add to the spectacular beauty of this area. Ocean winds ensure a mild climate, making the Lake Country a popular tourist destination.

## The stormy south

The trees of the Lake Country gradually become smaller and eventually disappear in the bitter, damp cold of the Archipelago. This region, which extends for about 1,000 miles (1,600 kilometers) from the Central Valley to Cape Horn, is a wild, windswept area of dense forests, snow-covered glaciers, and icy lakes.

Cold rains, piercing winds, and violent storms frequently pound the coastline of the Archipelago. The region's western edge is broken into thousands of tiny islands. At the southern tip of Chile lie the islands of *Tierra del Fuego* (Land of Fire). They were named by the famous Portuguese explorer Ferdinand Magellan in 1520 for the fires that Indians lit on the shores to keep themselves warm. Chile and Argentina each own some of the islands of Tierra del Fuego.

Chile also governs the Juan Fernández Islands, situated 400 miles (640 kilometers) off the mainland's west coast. Of the three islands in this group, Róbinson Crusoe Island is perhaps the most famous. A Scottish sailor named Alexander Selkirk lived alone on the island for more than four years between 1704 and 1709, and his adventures became the basis for Daniel Defoe's novel *Robinson Crusoe*. About 2,300 miles (3,700 kilometers) westward in the South Pacific, huge stone statues carved hundreds of years ago still stand on Easter Island's volcanic slopes.

Chile's narrow breadth extends from just north of the Tropic of Capricorn to the icy waters of Cape Horn. Chile is sandwiched between the towering Andes Mountains in the east and the Pacific Ocean on the west.

A geyser fed by an underground spring spurts hot water and steam above the barren landscape of the Atacama Desert. The only river that crosses the Atacama Desert is the Loa. It flows from the Andes Mountains to the Pacific Ocean through one of the world's driest areas.

# PEOPLE

Much of Chile remains uninhabited because of its harsh environment. Relatively few people live in the Northern Desert and the Archipelago. Most Chileans make their home in the Central Valley, where the climate is pleasant and the soil is rich. Nearly 90 percent of all Chileans live in cities and towns. The rest live in rural areas.

About 65 percent of Chile's people are *mestizos*—people of mixed Spanish and Indian ancestry. About 35 percent of the population is of unmixed European descent, mostly Spanish. About 7 percent of the people are of unmixed Indian ancestry.

Merchants in a Santiago seafood market offer the day's fresh catch. Chile's Pacific waters, particularly off the north coast, are rich in anchovettas, jack mackerel, and sardines. Chile's fishing industry is one of the largest in the world. Most of the fish are processed into fish meal and fish oil for export. Chileans also enjoy eating fish in such dishes as empanadas—stuffed turnovers.

Past and present meet in Santiago's Plaza de Armas, where a modern office block towers above a cathedral built in the Spanish colonial style. The capital and largest city in Chile, Santiago is also the nation's cultural, economic, and transportation center.

## Ethnic groups

About 1 million Mapuche—descendants of the warriors who defended their land against the Inca and the Spanish conquistadors—form the largest *indigenous* (native Indian) group in Chile. During the 1800's, the Mapuche controlled huge areas of the Lake Country. Today, most indigenous Chileans live in urban areas, but about one-fifth of the Mapuche still live in rural areas of southern Chile

The indigenous population also includes the Aymara, Quechua, and Atacama, some of whom live in oases in the Northern Desert region. The Aymara and Quechua probably migrated to the region from Bolivia and Peru in the 1900's. A few smaller groups are nomads roaming the Archipelago. Before the Spanish conquest, many other Indian tribes lived in the Archipelago, but they either were killed by the invading Spanish armies or died of diseases brought by the Spaniards, to which they had no natural resistance.

## City life

Much of the indigenous population exists outside the mainstream of Chilean society. Social classes in Chile are based chiefly on wealth, not ancestry. But nearly all members of the small, rich upper class are of European descent. Mestizos make up most of the middle class. The lower class consists mainly of poor mestizos and most of Chile's Indians.

Chile's lower class is concentrated in the crowded *callampas* (shantytowns) surrounding Santiago and other large cities. These slums are called *callampas*— the Spanish word for *mushrooms*—because they seem to spring up overnight.

The people who live in the callampas have moved to the cities from rural areas in search of higher-paying jobs and a better life. However, there is not enough work or low-income housing to meet the needs of the huge population.

The small Chilean upper class, on the other hand, lives in luxurious high-rise apartment buildings or spacious houses with well-kept lawns and gardens. Members of the middle class—whose ranks include government, industrial, and professional workers—also enjoy comfortable housing in modern apartments or single-family homes.

## Rural people

Most of the Chileans who remain in the rural areas are farmers. They own small farms or work as laborers or sharecroppers on large farms. Although conditions have improved somewhat since the 1960's, life is still difficult for these farmers, who barely make enough to support their families.

Before the 1960's, most poor Chileans worked as *inquilinos* (farmworkers) on huge estates called *fundos*, which were owned by a few wealthy families. In exchange for their labor, the inquilinos received a small plot of land to farm for themselves, but these plots were often too small to provide enough food for a family. In the 1960's, the Chilean government divided up many huge estates, enabling some inquilinos to own their own small farms.

A shepherd displays newly sheared wool. Sheep production is particularly high in the cold, wet climate of southern Chile, where the sheep grow a thick, heavy wool coat.

Santiago is Chile's capital and largest city. The city lies in the fertile Central Valley, the agricultural heartland of the country. The snow-covered Andes Mountains rise to the east of Santiago.

# ECONOMY

High in the mountains of the Atacama Desert, at an altitude of 9,900 feet (3,000 meters), the Chuquicamata mine gouges a huge scar in the earth. Chuquicamata is one of the largest open-pit copper mines in the world. Southeast of Santiago, in the Central Valley, lies El Teniente, the world's largest underground copper mine. Along with Chuquicamata and other mines, El Teniente produces large amounts of the reddish ore that Chile depends on for so much of its export.

With about a third of the world's known copper reserves, Chile is the world's leading copper-producing nation. The nation's mineral wealth also includes coal, gold, iron ore, lead, lithium, manganese, molybdenum, petroleum, and silver.

Though Chile has benefited greatly from its minerals, too much dependence on the profits from copper and other mining interests is dangerous to the nation's overall economy. Since the Great Depression of the 1930's, when the market for copper almost completely dried up, Chile has tried to develop other sources of revenue.

## Service industries and manufacturing

Today, service industries account for about half of the *gross domestic product*—the value of all goods and services produced yearly within the country. Many of Chile's service workers are employed by stores, restaurants, hotels, banks, health care facilities, social service organizations, and government agencies. Others are involved in transportation, communication, and such professions as teaching and law.

Manufacturing also contributes to the nation's economy, and Concepción, Santiago, and Valparaíso are the leading industrial centers. Chile's plants and factories produce a variety of consumer goods, including beverages, clothing, processed foods, textiles, and wood products. Other manufactured goods include cement, chemicals, copper, steel, and transportation equipment.

Copper is Chile's most valuable resource and export. Many other industries of Chile are dependent on the country's mineral production. Because minerals have long made up a large percentage of Chile's exports, the nation's economic health has been closely linked to the world market price for copper and other metals.

Chile's rich store of minerals includes large deposits of copper and molybdenum. Other imortant mining products include gold, iodine, iron ore, and natural sodium nitrate. Minerals account for nearly half of the total value of Chile's exports.

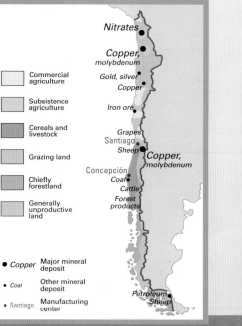

Commercial agriculture

Subsistence agriculture

Cereals and livestock

Grazing land

Chiefly forestland

Generally unproductive land

● Copper — Major mineral deposit

● Coal — Other mineral deposit

● Santiago — Manufacturing center

Nitrates
Copper, molybdenum
Gold, silver
Copper
Iron ore
Grapes
Santiago
Sheep
Copper, molybdenum
Concepción
Coal
Cattle
Forest products
Petroleum
Sheep

In copper production, ore is first ground into small pieces (1). Then water is added to the crushed ore, and the mixture, called slurry, passes into ball mills (2), which rotate drum-shaped cylinders half filled with iron balls. The slurry next goes into flotation cells, which concentrate the mineral-bearing particles. Smelting (4) removes most of the remaining impurities. The new mixture, called copper matte, then goes through a converter (5), where blowers force air through it and silica is added to it. The mixture is readied for casting (6) or electrolytic refining.

A lumber worker in Chile's forest industry saws through logs. About a third of the land in Chile is covered with forests. Timber, paper, pulp, and other wood products hold great promise for Chile's economy.

Grapes grow in the fertile Casablanca Valley. These grapes are used to make merlot wine. Chile is one of the world's largest producers and exporters of wine.

## Agriculture

The mild climate and fertile soil of the Central Valley provide good growing conditions, and the farmers in this region grow barley, corn, oats, rice, and wheat. Beans, potatoes, sugar beets, tomatoes, and other vegetables, as well as apples, citrus fruits, peaches, and wine and table grapes, are also grown in the Central Valley.

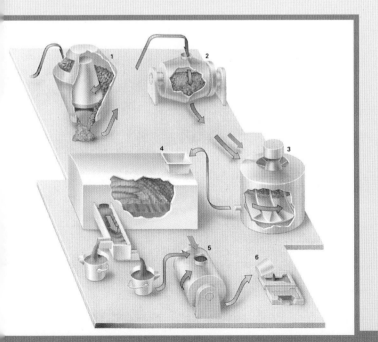

The harsh climate and rugged terrain of Chile's other land regions make agriculture less profitable. Only 3 percent of the land in Chile can be cultivated. Chile's farmers are unable to grow enough food to feed all its people, and the nation must rely on food imports.

Chile's land policies also limit agricultural production. In the 1960's, when the government began to break up the huge fundos and redistribute the land among poor farmers, many of the new plots were barely large enough to support a family. Land reform ended after the military took control of the government in 1973, and it has not resumed.

Today, most land is owned by small farmers or large corporations. Most of Chile's farms cover less than 25 acres (10 hectares). Small farmers are often too poor to purchase modern technology to work their land, but corporate farm production is increasing. Many corporate farms grow fruits and vegetables that are exported to Europe and North America during their winters.

# CHINA

China, a huge country in eastern Asia, is the world's third largest nation in area and the largest in population. About 20 percent of the world's people live in China.

This vast country is marked by an extraordinary variety of climates and landscapes. China's climatic regions range from tropical conditions in the southeast to severe, freezing weather in the north. Its landscapes range from some of the world's driest deserts and highest mountains to some of its richest farmland. The beautiful limestone hills near the city of Guilin on the Li River are among the most unusual features of this enormous country.

China is one of the world's oldest civilizations, with a written history that goes back about 3,500 years. The Chinese were the first people to develop the compass, gunpowder, movable type, paper, porcelain, and silk cloth. Numerous schools of philosophy and religion, including Confucianism and Taoism, began in China. The Chinese also built great cities and created magnificent works of art. Over the centuries, Japan, Korea, and other Asian lands have borrowed from Chinese art, language, literature, religion, government, and technology.

The Chinese call their country *Zhongguo,* which means *Middle Country* and probably originated from the ancient Chinese belief that their country was the geographical and cultural center of the world. The name *China* was given to the country by foreigners and may have come from *Qin* (pronounced *chihn*), the name of an early Chinese *dynasty* (series of rulers).

Today, China's official name is *Zhonghua Renmin Gongheguo* (People's Republic of China). The Chinese Communist Party gave the country this name in 1949 when they set up China's present government.

Since the Communists came to power, China has undergone many major changes. For example, most of the country's industries, trade, and finance have been placed under government control in an effort to make the nation an industrial power. As a result, the Communists have dramatically increased industrial production and expanded and improved education and medical care.

In the late 1900's, the Communists began to loosen their grip on the nation's economy and to allow more free enterprise. Today, China has one of the world's largest economies. Many of its people prosper, but the majority of Chinese live modestly. China has some of the world's largest cities, yet about half of its people still live in rural villages. Agriculture employs nearly half of all Chinese workers.

# CHINA TODAY

Since the 1980's, China has tried to move toward an open-market economy while retaining its Communist political system. Chinese leader Deng Xiaoping was credited with decentralizing economic decisions and introducing "socialism with Chinese characteristics." The decentralization led to great economic expansion. The number of privately owned and operated businesses has increased dramatically in China. Many experts believe the increased ownership of business has contributed significantly to China's economic growth.

The vibrant economy created much prosperity, but it did so unevenly. In 1994, Premier Li Peng warned of the danger of public disorder stemming from corruption and the disparities of wealth. He told China's parliament that the country needed to fight "money-worship, ultraindividualism, and decadent lifestyles." The economic changes displaced millions of rural workers, who moved around the country seeking work. With so many available workers, entrepreneurs could keep wages low. Meanwhile, inflation was high, and crime, especially corruption, was increasing.

Tensions arose as China's leaders allowed economic autonomy but retained political control. These tensions came to a head on April 18, 1989, when about 2,000 students rallied in Tiananmen Square, a huge open area in the center of Beijing next to the former imperial residence, the Forbidden City. There, the students held pro-democracy demonstrations. For six weeks, thousands of students set up camp in the square, boycotted classes, and continued the demonstrations.

Early in the morning of June 4, government soldiers entered the square, shooting at students and setting fire to their tents. Occasional shootings continued for several days as the troops established control, killing hundreds of protestors.

After the massacre, the government arrested thousands of people suspected of being involved

## FACTS

| | |
|---|---|
| Official name: | Zhonghua Renmin Gongheguo (People's Republic of China) |
| Capital: | Beijing |
| Terrain: | Mostly mountains, high plateaus, deserts in west; plains, deltas, and hills in east |
| Area: | 3,691,943 mi² (9,562,088 km²) |
| Climate: | Extremely diverse; tropical in south to subarctic in north |
| Main rivers: | Huang He, Yangtze, Xi Jiang |
| Highest elevation: | Mount Everest, 29,035 ft (8,850 m) |
| Lowest elevation: | Turpan Pendi, 505 ft (154 m) below sea level |
| Form of government: | Communist state |
| Head of state: | President |
| Head of government: | Premier |
| Administrative areas: | 22 sheng (provinces), 5 zizhiqu (autonomous regions), 4 shi (municipalities), 2 special administrative regions |
| Legislature: | Quanguo Renmin Daibiao Dahui (National People's Congress) with about 3,000 members serving five-year terms |
| Court system: | Supreme People's Court |
| Armed forces: | 2,185,000 troops |
| National holiday: | Anniversary of the founding of the People's Republic of China - October 1 (1949) |
| Estimated 2010 population: | 1,355,350,000 |
| Population density: | 367 persons per mi² (142 per km²) |
| Population distribution: | 55% rural, 45% urban |
| Life expectancy in years: | Male, 71; female, 75 |
| Doctors per 1,000 people: | 1.4 |
| Birth rate per 1,000: | 12 |
| Death rate per 1,000: | 7 |
| Infant mortality: | 20 deaths per 1,000 live births |
| Age structure: | 0-14: 20%; 15-64: 72%; 65 and over: 8% |
| Internet users per 100 people: | 22 |
| Internet code: | .cn |
| Languages spoken: | Mandarin Chinese (Putonghua), Northern Min (spoken in northern Fujian province), Southern Min (spoken mainly in Guangdong, Hainan, and southern Fujian), Wu (spoken in Shanghai, Jiangsu, and Zhejiang), and Yue or Cantonese (spoken in Guangdong and Guangxi) |
| Religions: | Atheist (official), Taoist, Buddhist, Christian, Muslims |
| Currency: | Yuan |
| Gross domestic product (GDP) in 2008: | $4.161 trillion U.S. |
| Real annual growth rate (2008): | 9.8% |
| GDP per capita (2008): | $3,090 U.S. |
| Goods exported: | Clothing, electronic devices, food, furniture, textiles, toys |
| Goods imported: | Chemicals, machinery, metals, petroleum |
| Trading partners: | Germany, Japan, South Korea, United States |

The People's Republic of China covers more than a fifth of Asia. It has a population of more than 1.3 billion people.

in the demonstrations. Some were tried and executed immediately. The Chinese government then instituted repressive controls within the country. China's violent suppression of the pro-democracy movement was denounced by most other nations.

In 1997, an agreement between China and the United Kingdom made Hong Kong a special administrative region of China. The United Kingdom had controlled Hong Kong since the mid-1800's. A similar agreement between China and Portugal returned Macau to Chinese administration in 1999.

In 2001, China became a member of the World Trade Organization, which promotes trade among its members. Membership marked progress in freeing the Chinese economy from government control.

Since the 1990's, China has spent huge amounts of money to improve its *infrastructure* (roads, bridges, dams, and other public works). These investments included major construction projects to improve facilities and services in Beijing in time for the city to host the Summer Olympic Games in August 2008.

# IMPERIAL CHINA

The first distinctly Chinese civilization emerged from two cultures—the Yangshao and the Longshan—that had developed by about 10,000 B.C. in what is now northern China. Many scholars regard the Xia culture, which probably arose during the 2100's B.C., as China's first dynasty.

The Shang dynasty arose from the Longshan and Xia cultures about 1766 B.C. Around 1045 B.C., the Zhou people of western China overthrew the Shang and ruled until 256 B.C. During the early Zhou period, about 500 B.C., the philosopher Confucius developed a system of moral standards that influenced Chinese society for more than 2,000 years.

## Early empires

In 221 B.C., the Qin state defeated the last of several rival states that were struggling for control of China. Although the Qin dynasty only lasted until 206 B.C., it founded the first Chinese empire controlled by a strong central government.

After the Qin dynasty collapsed, the Han dynasty gained control of China in 206 B.C. and ruled until A.D. 220. Arts and sciences thrived during this period, and Buddhism was introduced into China from India. A rebellion in 220 ended the Han dynasty, and China split into three rival kingdoms. The Sui dynasty reunified China in the 580's.

During the Tang dynasty, which replaced the Sui in 618, China enjoyed nearly 300 years of prosperity and cultural accomplishment. Another rebellion led to the collapse of the Tang empire in 907, and many dynasties fought for control of the shattered empire. The Song dynasty reunified China in 960. It supported the development of *Neo-Confucianism,* which combined Confucianism with elements of Buddhism and Taoism, and made it the state philosophy.

During the 1200's, Mongol armies invaded China from the north. In 1279, the Mongol leader Kublai Khan established the Yuan dynasty and made China part of the vast Mongol Empire, which extended over most of Asia. Marco Polo, a trader from Venice, lived in China from 1275 to 1292 and published an account of his experiences after he returned home.

The Mongols were driven out of China in the mid-1300's. The authoritarian Ming dynasty, established in 1368, ushered in a period of stability, prosperity, and flourishing arts that lasted until 1644.

## Manchu rule and Western powers

In 1644, the Manchu people of Manchuria invaded China and founded the Qing dynasty. The empire enjoyed stability until 1796, when political corruption touched off a rebellion that lasted until 1804 and weakened the Qing dynasty.

By the early 1800's, China exported large amounts of tea and silk to Europe, but bought little in return. To balance their trade, European merchants began to smuggle opium into China. When the Chinese tried to stop the illegal trade, the Opium War broke out between China and the United Kingdom. China lost the war and, in 1842, signed the Treaty of Nanjing. The treaty gave Hong Kong to the United Kingdom and opened four other Chinese

*The Qin dynasty,* in 221 B.C. established China's first empire. *The Han Dynasty* gained control in 206 B.C. and expanded the empire into Central Asia. *The Qing Dynasty,* also known as the *Manchu Dynasty,* was the last ruling Chinese dynasty, lasting from 1644 to 1912.

# TIMELINE

| | |
|---|---|
| c. 600,000-c. 200,000 B.C. | Prehistoric human beings known as the Peking (Beijing) people lived in northern China. |
| c 10,000-c. 1700's B.C. | New Stone Age cultures, including Yangshao and Longshan, flourish. |
| c. 1766-c. 1045 B.C. | The Shang dynasty rules China. |
| c. 1045 B.C. | The Zhou overthrow the Shang and establish new dynasty that rules until 256 B.C. |
| c. 500 B.C. | Philosopher Confucius develops system of moral values. |
| 221-206 B.C. | Qin dynasty establishes China's first strong central government and begins the Great Wall. |
| 206 B.C.-A.D. 220 | China becomes powerful empire under Han dynasty. |
| A.D. 220-581 | China divided, Buddhism spreads. |
| 581-618 | Sui dynasty reunifies China. |
| 618-907 | Tang dynasty rules China during period of prosperity and cultural achievement. |
| 960-1279 | Compass and movable type invented under Song dynasty. |
| 1126 | Rival dynasty from northeastern China conquers north. |
| 1275-1292 | Marco Polo visits China. |
| 1279-1368 | Mongols rule China under Yuan dynasty. |
| 1368-1644 | The Ming dynasty governs China. |
| 1644 | The Manchus establish Qing dynasty and rule for 268 years. |
| 1842 | Treaty of Nanjing gives Hong Kong to the United Kingdom and opens five other ports to British trade. |
| 1850-1864 | Millions die during the Taiping Rebellion. |
| 1894-1895 | Japan defeats China and gains control of Korea and Taiwan. |
| 1900 | Boxer Rebellion—an anti-Western campaign waged by Chinese secret societies—crushed by troops from eight nations. |
| 1912 | Republic of China established. |

**Confucius (551?-479? B.C.)**

**Kublai Khan (1215-1294)**

**Qin Shi Huang (259?-210 B.C.)**

ports to British trade. During the 1800's, China signed similar treaties with other Western nations.

From 1850 to 1864, the Taiping Rebellion threatened the survival of the Qing dynasty. The rebellion was put down, but millions of people were killed. Then China was defeated in a war with Japan in 1894 and 1895, further weakening the country.

In 1899, in order to prevent a single Western power from becoming dominant in China, the United States established the Open-Door Policy, which guaranteed the rights of all nations to trade with China on an equal basis. Some Chinese opposed the spread of Western influences in China, and in 1900 a secret society called the Boxers led a rebellion in which Westerners and Chinese Christians were attacked and killed. Troops from eight nations crushed the rebellion.

The Manchus tried to reform the Chinese government and economy, but these reforms did not save the dynasty. In the early 1900's, a movement to set up a republic began to grow, led by a Western-educated physician named Sun Yat-sen. The revolution that began on Oct. 11, 1911, eventually overthrew Manchu rule and established the Republic of China.

Beijing's Forbidden City—so called because only the emperor's household could enter it—includes the palaces of former Chinese emperors. The Forbidden City was built during the 1440's on the site of the palace of Mongol leader Kublai Khan.

# MODERN HISTORY

In 1912, the republican revolution forced the last Manchu emperor—a 6-year-old boy named Pu Yi—from China's throne. Yuan Shikai became president of the Republic of China in place of Sun Yat-sen, who agreed to step down, but Yuan's presidency soon became a dictatorship. In 1913, the republicans established the Kuomintang (Nationalist Party) and tried to overthrow Yuan. The revolt failed, however, and the Kuomintang leaders fled to Japan.

When Yuan died in 1916, the central government weakened, and the real power in northern China passed into the hands of *war lords* (local military leaders). Sun Yat-sen set up a rival government in Guangzhou in southern China, but by 1922 his republic had failed and civil war was widespread.

In 1919, Sun had begun to reorganize the Kuomintang. The first Communist student groups also appeared at this time. In 1923, the Soviet Union sent advisers to aid the Nationalists. The Soviets convinced the Chinese Communists to work with the Nationalists.

After Sun Yat-sen died in 1925, the Nationalist Party was led by its military commander, Chiang Kai-shek. In 1926, Chiang set out to reunite China, but tensions developed between the Nationalists and the Communists. In 1927, Chiang and his troops turned against the Communists, who retreated to the hills in southern China. The Nationalists captured Beijing and united China under one government in 1928.

**Yuan Shikai
(1859-1916)**

**Chiang Kai-shek
(1887-1975)**

**Zhou Enlai
(1898-1976)**

## Nationalist rule

The Nationalist government was plagued from the start by Communist opposition and Japanese aggression. Then, in 1931, the Japanese seized Manchuria and extended their military influence into other parts of northern China. In 1934, Chiang's armies forced the Communists to evacuate their bases in southern and central China, but they established a new base in Shaanxi in northern China, after marching more than 6,000 miles (9,700 kilometers) in the famous *Long March*. During that fateful journey, Mao Zedong became the leader of the Chinese Communist Party.

In 1937, the Japanese launched a major attack against China, and the Nationalists and Communists formed a united front to oppose them. In World War II (1939-1945), China fought with the Allies against Japan. By the end of that war in 1945, the Chinese Communists held an area in northern China and commanded a large army.

In 1946, a full-scale civil war broke out between the Nationalists and the Communists. The superior military tactics of the Communists and the social revolution they conducted in the countryside gradually turned the tide against the Nationalists. Mao Zedong established the People's Republic of China on Oct. 1,1949. In December, Chiang and his followers fled to Taiwan.

- Boundary of present-day China
- Chinese Communist areas
- Japanese controlled areas

U.S.S.R.

Mongolia

Manchuria

Korea

**CHINA**

Yan'an

Japan

Tibet

The Long March
1934-1935

Nanjing

India

Burma

Indochina

In 1931, Japan seized control of Manchuria. In 1934, the Nationalist government forced the Communists to evacuate their bases in southern and central China, but the Communists established a new base in northern China after the famous *Long March* of 1934 and 1935.

# TIMELINE

Mao Zedong proclaims the founding of the People's Republic of China in Beijing on Oct. 1, 1949. Mao gave up the title of chairman of the People's Republic in 1959 but remained in control of the country and of the Communist Party until his death in 1976.

## Communist rule

The Communists, under Party Chairman Mao Zedong, initiated an economic plan based on state control and the development of heavy industry. The new government also redistributed farmland among the peasants and combined the landholdings into large cooperatives.

In 1958, the Communists launched the Great Leap Forward, a campaign designed to accelerate economic development. The program shattered China's economy, however, resulting in food shortages and a decline in industrial production.

In 1966, Mao launched the Cultural Revolution, another disastrous campaign aimed at preserving Communist principles. During the Cultural Revolution, radicals warred against moderate party members, and China came close to civil war.

When Mao Zedong died in 1976, the moderates came to power and liberalized some of Mao's strict policies. In 1979, the new leaders established normal diplomatic relations with the United States.

By 1980, Deng Xiaoping, a moderate, had become China's leader. He began economic reforms and increased cultural contact with Western nations, but in 1989 many students demonstrated for increased freedoms. Chinese soldiers attacked the students gathered in Beijing's Tiananmen Square and killed hundreds. Although Deng Xiaoping resigned later that year, he remained China's most influential leader until his death in 1997.

Thousands of Chinese students and workers demonstrate in support of democratic reforms in 1989 at Beijing's Tiananmen Square and in other Chinese cities. Their bold challenge to the country's Communist leaders brought a brutal military crackdown.

# CHINESE CIVILIZATION

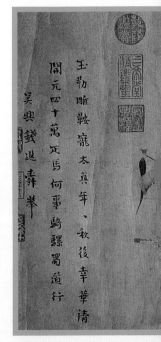

China has one of the world's oldest living civilizations, and many other countries have benefited from Chinese technology, philosophy, and culture over the centuries. About 1,000 years ago, Chinese scientists invented the compass. Movable type for printing was also invented in China during this period—approximately 400 years before its development in Europe. Many Asian lands adopted Confucian and Taoist values.

Chinese civilization has been heavily influenced by Confucianism, Taoism, and Buddhism. Confucianism offered a set of moral values stressing order, discipline, and rules of behavior that benefited society. Taoism, on the other hand, dealt with more personal problems, such as a person's ability to live in harmony with nature. Buddhism, which originated in India, encouraged spiritual thinking, right living, and detachment from worldly things. China's literature, art, and music reflect all these traditional religions and values.

## Sculpture and pottery

Chinese pottery and jade from the 4000's B.C. showed great technical skill and artistic refinement. Chinese potters started to use the pottery wheel before 3000 B.C. Bronze urns and vases were used before the 1500's B.C. in religious ceremonies that included ancestor worship.

Excavations of tombs and dwellings have yielded many other ancient works of Chinese art. In 1974, for example, archaeologists uncovered thousands of life-sized clay figures of soldiers and horses in burial pits near the city of Xi'an. These figures stood guard over the tomb of emperor Shi Huangdi, who founded the Qin dynasty in the 200's B.C. In March 1990, about 25 miles (40 kilometers) from that site, a road crew discovered a network of vaults that extend over an area about the size of 10 football fields and contain tens of thousands of terra-cotta sculptures of men, boys, horses, and carts.

After Buddhism reached China from India during the Han dynasty (206 B.C.-A.D. 220), sculptors

Porcelain, the most delicate type of pottery, was developed by the Chinese in the A.D. 100's. Fine porcelain dishes and vases are among the greatest treasures of Chinese art.

carved images of Buddha to decorate the temples. Some of these sculptures were made of stone or clay, while others were cast in bronze and coated with gold. In rural areas, elaborate Buddhist chapels were hollowed out of cliffsides and decorated with figures of Buddha.

## Painting, literature, and drama

As early as the 4000's B.C., Chinese potters decorated their works with sophisticated painted designs. Painting on silk began during the Shang dynasty (about 1766-about 1045 B.C.). Painting on paper began after the Chinese invented paper in the 200's B.C. Most of the early paintings show people, gods, or spirits, but by the A.D. 900's, many artists painted landscapes. In these paintings, called *shanshui* (mountain-water), artists used towering mountains and vast expanses of water to communicate a feeling of Taoist-inspired harmony between nature and the human spirit.

In China, *calligraphy* (fine handwriting) is a branch of painting and considered an art form. Beginning in the 1000's, an

A clay army of about 10,000 terra-cotta soldiers and horses was excavated near Xi'an in 1974. In the 1100's B.C., the soldiers and servants of a ruler were buried alive in their master's tomb when he died. Over time that custom was modified, and clay images were substituted for living people.

A scroll from the Tang dynasty shows a court scene in ancient China. The artist has used calligraphy to enhance the beauty of the painting and describe the circumstances under which it was created.

inscription or poem in exquisite calligraphy often formed part of a painting's overall design.

Chinese literature dates back almost 3,000 years and includes many great works. Some, inspired by Confucianism, teach moral lessons, while poems that celebrate the beauties of nature reflect the influence of Taoism. One of China's greatest poets, Wang Wei, lived during the Tang dynasty (A.D. 618–907). His works, which emphasize quietness and contemplation, demonstrate the influence of Buddhism.

Before the 1900's, almost all of China's greatest writers were given important government jobs, due to their skill with words. For this reason, many masterpieces of Chinese literature deal with history, politics, and science.

Chinese drama developed around the 1200's. However, since the 1800's, the most popular form of Chinese drama has been *Beijing opera*—plays based on Chinese history and folklore. Dressed in elaborate and colorful costumes, the actors of the Beijing opera combine dialogue and songs with dance and symbolic gestures.

A huge statue of Buddha, watches over a cliffside chapel in the Yunggang Grotto near Datong. More than 51,000 statues of Buddha and his attendants grace this magnificent grotto.

In traditional Chinese houses, large tile roofs with elegant, curved edges are supported by wooden columns. The design and location of Chinese buildings express the ideal of harmony between people and nature.

1 Entrance door
2 Living rooms divided by light partitions
3 Galleries link rooms
4 Wooden support posts
5 Tiled roof
6 Courtyard

# THE GREAT WALL

The Great Wall of China is the only manmade object visible from the moon and is the longest structure ever built. It stretches more than 5,000 miles (8,000 kilometers), including about 3,890 miles (6,260 kilometers) of handmade walls as well as such natural barriers as rivers and hills. The Chinese built the wall to protect their northern border from invasion. In times of war, troops and horses could march quickly along the wall, avoiding difficult mountain passes. Today, the Great Wall is truly one of the treasures of world civilization.

The Great Wall begins in the east near the North Korean border and winds its way to Jiayuguan in the west. Watchtowers, placed every 100 to 200 yards (91 to 180 meters) along the wall, once served as lookout posts. Although parts of the wall have crumbled through the years, much of it remains, and some sections have been restored.

In the east, the wall winds through a mountainous region called the Inner Mongolian Border Uplands. Granite blocks form the foundation of this part of the Great Wall, while the sides are made of stone or brick. The inside of the wall is filled with earth. Layers of brick placed on top of the earth form a roadway.

Farther west, the Great Wall runs through hilly areas and along the borders of deserts. Stone and brick were scarce in these areas, so the workers built this section of the wall out of earth, moistened and pounded to make it solid.

The first sections of what was to become the Great Wall of China were probably built during the 400's B.C. by the peoples of small, warring states. Some states constructed long walls of packed earth to serve as borders marking their territory.

The Great Wall of China is the longest structure ever built. Its handmade walls stretch about 5,500 miles (6,260 kilometers) over such natural barriers as rivers and hills. The Great Wall extends from the Bo Gulf of the Yellow Sea to the Jiayu Pass in what is now Gansu province in western China.

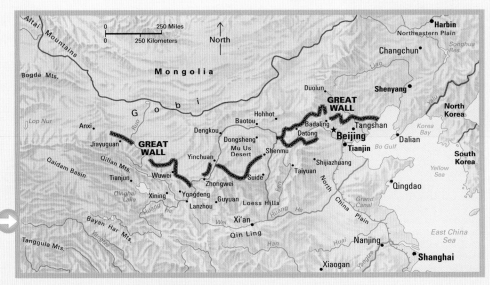

The Great Wall of China represents one of the world's great engineering feats. The wall stands about 25 feet (7.6 meters) high.

The Qin dynasty (221-206 B.C.) conquered these smaller states. Shi Huangdi, the first Qin emperor, is regarded as the first ruler to conceive of and build a Great Wall. He had new walls built to connect the older ones. The Great Wall was constructed at huge expense—and at the cost of many lives. Thousands of workers toiled for years to build it.

Work on the wall continued through the Han dynasty (206 B.C.-A.D. 220) and the Sui dynasty (581-618), but much of the wall gradually fell into ruin after these periods. The structure continued to provide protection against minor attacks, but in the 1200's, Mongol warriors led by Kublai Khan swept across the wall and conquered much of China.

The Great Wall was rebuilt during the Ming dynasty (1368-1644), which followed the Mongol rule. To prevent further invasion from hostile foreigners, the Ming emperors had soldiers on guard in the watchtowers at all times. At particularly vulnerable points, several walls were built—one behind the other—to strengthen the defenses. Almost all of the Great Wall as it appears today dates from the Ming period.

Over the centuries, much of the Great Wall again collapsed. Since the late 1900's, however, the Chinese government has done extensive renovations at the most visited sections. These sections are near the east coast, outside Beijing, and in the province of Gansu in north-central China. Near Beijing, a monumental gateway marks a restored section of the wall at Badaling, where the roadway atop the wall is wide enough for five horses to gallop side by side.

Today, the Chinese no longer depend upon the Great Wall to keep enemies out, but rather to bring visitors in. Thousands of tourists from China and other countries come to see the wall every year. The Great Wall is also of interest to historians, who study the structure's fortifications, and to scientists, who study the effects of earthquakes on the wall.

# WESTERN LAND REGIONS

China, the world's third largest country, spans about 2,500 miles (4,023 kilometers) from north to south, and approximately 3,000 miles (4,828 kilometers) from east to west. This vast land has a remarkably wide variety of landscapes and climates.

Western China is a land of high mountain ranges and vast areas of deserts and *steppes* (dry grasslands). Mountains form natural boundaries along much of China's southern and western border regions, which are sparsely populated. The Gobi, Mu Us, and Takli-makan deserts lie in northern and northwestern China. These areas are also thinly populated. Western China can be divided into two large land regions—the Ti-betan Highlands and the Xinjiang-Mongolian Uplands.

## The Tibetan Highlands

The Tibetan Highlands lie in southwestern China. The region contains the high, cold Plateau of Tibet, bor-dered by towering mountains—the Himalaya on the south, the Karakorum Range and the Pamirs on the west, and the Kunlun on the north. This region includes Tibet, which has been part of China since the 1950's.

A Kazakh campsite, with its warm felt tents and peaceful animals, makes a cozy scene on the lower slopes of the Altai Mountains in the Xinjiang-Mongolian Uplands. The Kazakh people, one of China's minority groups, are herders in northwestern China.

Farming villages in Qinghai province nestle in the mountain valleys of the Tibetan High-lands. Crops such as bar-ley and wheat can be grown in these sheltered valleys.

Tibet stands at an average elevation of 16,000 feet (4,880 meters), and the highest mountain in the world—Mount Everest— rises 29,035 feet (8,850 meters) in the Himalaya in southern Tibet. In the north, many peaks of the Kunlun range rise more than 20,000 feet (6,000 me-ters). Two of China's most important rivers, the Huang He and Yangtze, begin in the mountains of the Tibetan Highlands and flow eastward across China to the sea.

Tibet's harsh climate is characterized by drought and long, bitterly cold winters, while violent winds sweep the region the year around. Most of the Highlands is a wasteland of rock, gravel, snow, and ice, though a few areas provide limited grazing for hardy *yaks*—hairy oxen that furnish food, clothing, and transportation for the Tibetan people. Most of the land cannot support agriculture due to the poor soil and cold climate. However, crops can be grown in some fertile valleys in the south.

## Xinjiang-Mongolian Uplands

The Xinjiang-Mongolian Uplands cover the vast desert areas of northwestern China, and Xinjiang lies in the western part of the region. Although Xinjiang covers about 17 percent of China's land area, only about 1.5 percent of China's people live in the region. However, Xinjiang is an important region economically, with vast deposits of coal, iron ore, oil, and uranium.

The Tian Shan mountains divide the Xinjiang region into two areas—the Taklimakan Desert to the south and the Jungga Basin to the north. The Taklimakan is one of the world's driest deserts, receiving less than 4 inches (10 centimeters) of rain a year. Most of the people of this region live on or near natural or artificially created desert oases. The Turpan Depression, an oasis near the northern edge of the Taklimakan, is the lowest point in China, lying 505 feet (154 meters) below sea level. The remote Junggar Basin north of the Tian Shan stretches northward to the Altai Mountains along the Mongolian border.

The eastern part of the Xinjiang-Mongolian Uplands contains part of the Gobi Desert and the Mu Us Desert. The Gobi consists mainly of dry, rocky, or sandy soil surrounded by steppes. Summers are hot in the Gobi, and daytime temperatures can reach 113° F (45° C). In winter, the temperature can drop to −40° F (−40° C). The dry, sandy Mu Us Desert lies in a region known as Inner Mongolia, bordering the Loess Hills in the Inner Mongolian Border Uplands.

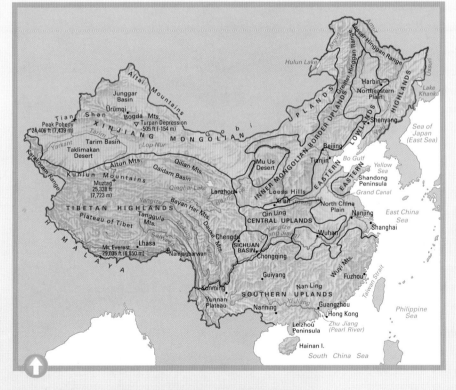

Western China has some of the highest mountains in the world. The land of Tibet in the Tibetan Highlands of southwestern China is often called the Roof of the World.

Camel caravans once transported silk and other goods on the Silk Road, an overland trade route from China to the West that crossed the forbidding wastes of the Taklimakan Desert.

# EASTERN LAND REGIONS

Eastern China contrasts sharply with the barren, thinly populated regions of western China. Most of China's people live crowded together in the eastern third of the country, which has nearly all the land suitable for farming and most of China's major cities.

Eastern China can be divided into six major land regions. In northeastern China, they are the Inner Mongolian Border Uplands, the Eastern Highlands, and the Eastern Lowlands. In southeastern China, they are the Central Uplands, the Sichuan Basin, and the Southern Uplands.

The Inner Mongolian Border Uplands lie between the Gobi Desert and the Eastern Lowlands, with the rugged Greater Hinggan Mountains to the north. Little agriculture is carried out in this area.

The southern part of the region is thickly covered with *loess,* a fine, yellowish soil of tiny mineral particles that is deposited by the wind and easily eroded. The *Huang He,* which means *Yellow River,* takes its name from the large amounts of loess it carries.

The Eastern Highlands consist of eastern Manchuria and the mountainous Shandong Peninsula, a region with excellent harbors and rich coal deposits. China's finest forestland covers the hills of eastern Manchuria, and timber is one of the region's major products.

The Eastern Lowlands consist of the Northeastern Plain, the North China Plain, and the Yangtze River Valley. It lies between the Inner Mongolian Border Uplands and the Eastern Highlands and extends south to the Southern Uplands. The Eastern Lowlands have China's most productive farmland and many of the country's largest cities, including Beijing and Shanghai.

The Northeastern Plain, sometimes called the Manchurian Plain, has fertile soils and large deposits of coal and iron ore. Farther south, wheat is the main crop in the rich agricultural area of the Huang He Valley in the North China Plain.

A rural village in the Sichuan Basin is surrounded by fertile hills. Land travel is difficult in the region, and rivers provide transportation routes for small boats.

The lower Yangtze Valley has the best combination of level land, fertile soil, and sufficient rainfall in China, and rice is the region's chief crop. In addition, the Yangtze River and its many tributaries are the country's most important trade routes.

The Central Uplands are an area of hills and mountains between the Eastern Lowlands and the Tibetan Highlands. The Qin Ling Mountains cross the region from east to west. They form a natural barricade against the seasonal monsoons that carry cold air and dust from the north in winter and carry warm air and rain from the south in summer. North of the mountains, the dry weather is beneficial in the cultivation of wheat, while rice is the major crop in the warm, humid areas to the south.

The city of Yan'an in the province of Shaanxi lies on a tributary of the Huang He in the fertile Loess Hills area of the Inner Mongolian Border Uplands. The Huang He and its tributaries have carved out hills and steep-sided valleys in the soft loess soil.

The Sichuan Basin, which lies south of the Central Uplands, is a region of hills and valleys surrounded by high mountains. With its mild climate and long growing season, the Sichuan Basin is one of China's major agricultural regions. The name *Sichuan* means *Four Rivers* and refers to the four streams that flow into the Yangtze in this region. The rivers have carved out deep gorges in the red sandstone, making land travel difficult.

The Southern Uplands, a region of green hills and mountains, cover southeastern China and include the island of Hainan. The deltas of the Xi Jiang (West River) and Min River are the only level areas and also the most densely populated parts of the region. Deep, rich soils and a tropical climate help make the deltas extremely productive agricultural regions. Guangzhou (also called Canton), southern China's largest city, lies near the mouth of the Xi Jiang.

Much of the Southern Uplands is so hilly and mountainous that farming is largely impossible. The Guilin Hills, one of the strangest and most beautiful sights in China, rise in the central part of the region.

The limestone hills near the city of Guilin owe their unusual shape to erosion and weathering. Many isolated limestone hills rise 100 to 600 feet (30 to 182 meters) almost straight up. Some people say the hills resemble rows of dragons' teeth.

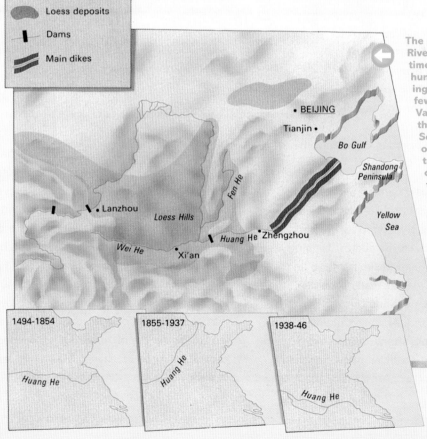

Loess deposits

Dams

Main dikes

BEIJING
Tianjin
Bo Gulf
Shandong Peninsula
Fen He
Lanzhou
Loess Hills
Yellow Sea
Wei He
Huang He · Zhengzhou
Xi'an

1494-1854
Huang He

1855-1937
Huang He

1938-46
Huang He

The course of the Huang He River has shifted several times during the last several hundred years. Major flooding formerly occurred every few years in the Huang He Valley, earning the river the nickname "China's Sorrow." Today, a system of dams and dikes controls most floods. Loess carried by the river fertilizes the North China Plain in the Eastern Lowlands.

# WILDLIFE

China's wildlife includes a great variety of animal species. The takin, a mammal that resembles a musk ox, inhabits the dense bamboo forests of central, western, and southwestern China, and the Himalaya. The Chinese alligator and the giant salamander live in the Yangtze River Valley. China is also home to many species of birds, including the rare Himalayan monal and Lady Amherst's pheasant—found only in the foothills of the Himalaya.

Many other rare and endangered animals also live in China. The giant panda, a black-and-white bearlike animal, lives only in western and southwestern China. Other rare animals include the snow leopard, which lives high in the mountains of central Asia, and several species of cranes. The Chinese dolphin, a small freshwater whale, is found in Dongting Lake in Hunan province.

## Endangered species

Many animal species in China are close to extinction, and in many cases their survival is threatened by the activities of human beings. For example, people hunt and kill the snow leopard for its beautiful fur. Similarly, Asian rhinoceroses are killed for their horns, and tigers in China have been hunted for food.

People have also seriously harmed and even destroyed many animal habitats to clear land for agriculture and industry. Marshes that serve as breeding grounds for cranes have been drained for farming and for settlements. In the Sichuan Basin, bamboo and rhododendron forests, as well as many broadleaf and coniferous trees, have been cut and burned to extend the region's agricultural land. Animals that live in the area—such as the antelope, shrew, golden monkey, and giant panda—depend on these plants for their food.

China's wildlife includes animals that live in high, mountainous regions, such as the yak (1) and the snow leopard (2). The golden monkey (3), red panda (4), Père David's deer (5), musk deer (6), giant panda (7), and tiger (8) all live in forested areas. Tiny shrews (9) and the giant salamander (10) make their home in and along China's rivers. The country's birds include such exotic specimens as the black-throated crane (11), the northern goshawk (12), the collared Scops owl (13), the Asian flycatcher (14), and Temminck's tragopan (15).

The destruction of bamboo forests has particularly threatened the survival of the giant panda, which feeds chiefly on bamboo shoots. However, the growth cycle of the bamboo plant has also contributed to its scarcity. Since the 1970's, some of the bamboo species eaten by the giant panda have been going through periodic die-offs. Every 15 to 120 years, bamboo plants flower, produce seeds, and then die. It takes several years for the seeds to grow into plants that can provide food for pandas. Such die-offs have led to the deaths of hundreds of pandas. Scientists estimate that fewer than 1,000 giant pandas remain in the wild.

## THE GIANT PANDA

The giant panda is one of the world's most endearing and popular animals. Zoologists long disagreed about how to classify the giant panda. Research conducted during the 1980's, however, revealed that the giant panda's chromosomes (the parts of a cell that carry genes) chemically resemble those of bears. As a result, most zoologists now consider the giant panda to be a bear. Panda cubs are extremely tiny, weighing only about 5 ounces (140 grams) at birth. Adults, on the other hand, weigh about 200 to 300 pounds (90 to 140 kilograms). A giant panda can grasp an object between its fingers and its so-called extra thumb. This thumb, which is actually a bone covered by a fleshy pad, grows from the wrist of each forepaw. Pandas also have true thumbs, which they use as fingers.

*The rare giant panda lives in bamboo forests on the upper mountain slopes of western and southwestern China.*

## Wildlife preservation

In an effort to save the remaining giant pandas and increase their population, the Chinese government has protected pandas by law. In addition, the government has loaned giant pandas to various zoos in the United States, hoping that the animals will breed in captivity. Unfortunately, most such attempts have failed.

By 2000, China had also set up hundreds of *nature reserves*—refuge areas where plants and animals are protected. The southern province of Yunnan is known as China's "plant and animal kingdom" because of its many protected areas.

Some of China's nature reserves protect an entire ecological system from people and pollution, conserving the natural balance between plants and animals and their habitat. Other areas have been designed to protect particular animal and plant species. For example, a small preserve near Dalian in northeastern China protects thousands of venomous snakes. In the past, many of these animals had been killed for their meat, which some Chinese consider a delicacy. In some nature reserves, trees such as the silver cypress are grown to replenish forests that have been cleared.

**Giant Panda Habitat**

- ■ Current range
- ■ Historic range

Beijing

Chengdu
Chongqing
Wuhan
Shanghai
Hong Kong

*Pandas currently occupy a much smaller range than they did hundreds of years ago. The cutting of forests for wood as well as for farmland and settlements has made most of their historic range unsuitable for pandas and the plants they feed on.*

# THE HAN

China has more than 1 billion people, and about 92 percent of them belong to the Han ethnic group, which has been the largest nationality in China for centuries. The Chinese call themselves Han people in recognition of China's achievements during the Han dynasty.

All the Han people speak Chinese, but spoken Chinese has many dialects that differ enough in pronunciation to be considered separate languages. About 70 percent of the nation's people speak Mandarin, or *putonghua* (common language), the country's official language. Other major dialects include Wu (spoken in Shanghai) and Yue (also called Cantonese). However, although the dialects differ in pronunciation, all the people who speak Chinese write the language in the same way.

## Changes under Communism

Family life has always been extremely important in Han Chinese culture, but family life has changed greatly since the Communists came to power in 1949. In the past, children were expected to obey their parents without question in all matters, and a father could legally kill his children if they disobeyed him. Parents also decided who their children would marry. Today, relationships within Chinese families are less formal and more flexible.

Before 1949, families valued sons far more than daughters, and girls were sometimes killed at birth to save resources for the family's sons. Today, the sexes are regarded more equally. Most women work outside the home, and many men share in the housekeeping.

The Communist government has also made great progress in education. Adult education classes are available for older people. In the mid-1980's, China's government began trying to get children to attend school for at least nine years. Education in China has alternated between supporting the Communist principle of equality and providing high-quality education for outstanding students.

## City and country life

In general, people in China's urban areas have a higher standard of living than people in the countryside. In the cities, most households have at least two wage earners, and the cost of food and rent are low. However, housing can be a problem in China's crowded cities. Many people live in new apartment complexes built by the government or by businesses for their workers. Houses in some older neighborhoods resemble those in the countryside.

A group of adults practices t'ai chi in the early morning at the Temple of Heaven in Beijing. T'ai chi is a form of Chinese martial arts chiefly used as a method of mental and physical relaxation.

Out for a stroll on a sunny day, a man takes care of his grandchildren while the children's parents work outside the home. Although this arrangement is widespread in China, a growing number of children attend nursery school and kindergarten.

More than half of China's people live in rural villages and small towns. Most rural families live in five- or six-room houses made of mud bricks, clay bricks, or stone, with roofs of tile or straw. Farmers work many hours a day and attend night classes and political meetings. Nevertheless, they still have time for recreation. Many villages have a small library and a recreation center with a television or a film projector, and sometimes a computer. Villages may also have sports facilities, musical groups, or theater groups.

City life offers a greater variety of cultural opportunities. People who live in cities have more classes to choose from. City dwellers also enjoy restaurants, parks, museums, theaters, and sporting events. Stores in the city provide a wider selection of foods and merchandise.

Many Han people—both city and rural dwellers—perform ancient Chinese exercises called *t'ai chi ch'uan* (also written *taijiquan*) every morning. *T'ai chi ch'uan* emphasizes relaxation, balance, and proper breathing techniques and is also a form of self-defense.

Most Chinese wear clothing similar to that of Europeans and North Americans. In urban areas, fashionable designs are popular, especially among younger people. In rural areas, people often make their own clothes.

China's population is concentrated mainly in the eastern half of the country. By law, to limit population increases, men may not marry until they are 22 years old, and women until they are 20. People are encouraged to have no more than two children.

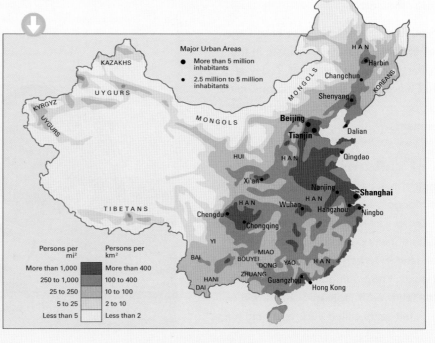

# MINORITY GROUPS

About 8 percent of China's people are members of minority groups. Kazakhs, Mongols, Tibetans, and Uyghurs are a few of the country's more than 50 ethnic minority groups. Most of China's minority peoples live in the border regions and in the western half of the country. Many of these ethnic groups retain their own way of life, culture, and language, in spite of government efforts to impose the ways of the Han Chinese on them.

Tensions between the minorities and the Han Chinese have sometimes led to open revolts. Many minority groups resent what they perceive as China's attempts to absorb their cultures, while many Han Chinese have a deep-seated prejudice against the minorities.

## Pressure to conform

The Chinese Communist government has tried to integrate the minority groups into the country's mainstream way of life. At times, this effort has included pressure to adopt practices of the majority, such as particular farming methods. Since 1976, however, policies toward the minority groups have become more moderate.

The Communists have also tried to encourage members of ethnic groups to participate in the Communist Party and in government. Today, although Han Chinese hold the key positions, minority members occupy some positions within their local governments.

## A clash of cultures

The government has had moderate success in integrating the Mongols who live in northern China—a region sometimes called *Inner Mongolia*. The Mongols have traditionally been nomadic herders, roaming the land in search of pastures for their sheep and goats. The traditional Mongolian customs are still followed in nomadic camps, where people live in felt tents called *yurts* and hold horse

Horse races and hunting are popular pastimes among the nomadic Kazakh people of northwestern China.

A Hani woman in a brightly colored head-dress shops at a market in Yunnan, along the border with Vietnam. The Hani people are among the many minority groups living in this part of southwestern China.

Mongol boys in Hohhot dress much like other schoolchildren in China. Mongols live along the Mongolian border but make up only a small percentage of Inner Mongolia's population.

races, archery contests, and wrestling matches. Today, however, many Mongol people have settled down on government-owned cooperative farms.

The government has had far less success in Tibet and Xinjiang, where minorities form the majority of the population. The Tibetan people, who live in China's southwestern highlands, have fiercely resisted Chinese rule, and their language, culture, and religion differ radically from those of the Han Chinese. Most Tibetans, who constitute much of the region's population, are barley farmers or nomadic herders. They resent the Chinese government's attempts to change their agricultural methods.

The Uyghurs and Kazakhs, who live in the deserts of northwestern China in the Xinjiang region, have also rebelled against the Chinese. The Uyghurs raise livestock and grow a wide variety of crops on oases in the deserts, while the Kazakhs herd sheep and goats. Both groups follow Islam—the major religion in the Middle East. Their language, clothing, architecture, and music also show Middle Eastern influence.

Many other minority groups live in the far southern parts of China. Although some of these groups speak Chinese dialects and live much like the Han Chinese, ethnic differences are strong among others. Many of these groups are related to the peoples of Myanmar, Laos, Vietnam, and Thailand. The people in several of these areas have adopted some aspects of Chinese culture, but more out of a fear of official displeasure than a fondness for the Han way of life.

For example, the T'ai people in the province of Yunnan, who are related to the people of Thailand, celebrate agricultural festivals and other special occasions with songs and dancing. However, during Mao Zedong's rule, dances that had been traditionally performed around harvest foods were done instead around a large picture of Mao.

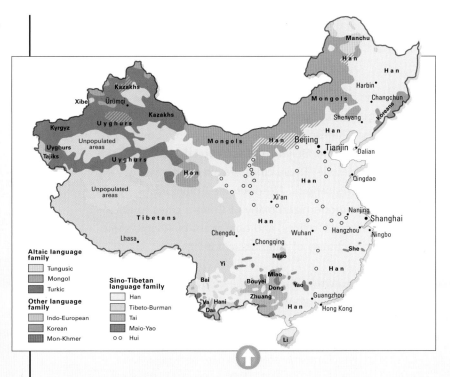

More than 50 minority groups, who live mainly in the border areas and in western China, make up only 8 percent of China's population. The remaining 92 percent belong to the Han ethnic group and live crowded together in eastern China.

Ethnic Uyghurs, who live primarily in the Xinjiang region in western China, are Muslim and ethnic Turkic people. They make up a large percentage of the region's population. The Chinese government has called Uyghur nationalists terrorists, and in July 2009, violent clashes between Uyghurs, Chinese police, and Han residents in the city of Ürümqi left nearly 200 people dead.

# CHINESE CITIES

More than a dozen of China's cities are among the 100 largest urban centers in the world. Shanghai is the largest Chinese metropolitan area, with more than 15 million people. Beijing, China's capital, has more than 11 million people. These cities are also important manufacturing centers. Although the Communist government modernized many of China's cities to make way for factories, offices, and apartment buildings, some still preserve traces of their ancient past.

## Xi'an

Xi'an lies in central China at the eastern end of the Silk Road, the great overland trade route that linked Rome to Chang'an (now called Xi'an). The ancient city of Chang'an, which means *eternal peace*, served as the capital of the Han (206 B.C.- A.D. 220), Sui (581-618), and Tang (618- 907) empires. Chang'an had more than 1 million residents during the Tang period, making it the largest city in the world at that time.

A street vendor sells soft drinks in the shadow of a statue of Mao Zedong in a busy square in Chengdu. Capital of Sichuan Province and a major industrial center in southern China, the city still has some of its traditional wooden houses and bustling markets.

The tomb of Sun Yat-sen lies on Zijin Mountain, east of Nanjing. Sun, a Nationalist leader, helped establish the Republic of China in 1912. When the Nationalists, under Chiang Kai-shek, united China under one government in 1928, they made Nanjing their capital.

Chang'an declined after the fall of the Tang dynasty, and attempts to restore the city during the Ming dynasty (1368-1644) were short-lived. Two Buddhist temples—the Little Wild Goose Pagoda and the Big Wild Goose Pagoda—are among the buildings that have survived from the Tang dynasty. The massive Bell Tower standing at the crossroads of Xi'an's two main streets was built during the Ming dynasty.

## Nanjing

People have lived in what is now the Nanjing area since about the 400's B.C. Nanjing, which means *southern capital*, was the seat of government during the first part of the Ming dynasty. It also served as the capital of the Republic of China from 1928 to 1937 and then again from 1946 to 1949. Today, the city is the capital of Jiangsu Province and an important center of industry, transportation, and government in east-central China.

Nanjing lies on the Yangtze River about 200 miles (320 kilometers) from the East China Sea. The wharves along the riverbank can handle oceangoing ships that sail up the river. A 4.2-mile (6.7-kilometer), double-deck bridge completed in 1968 takes trains and motor vehicles across the Yangtze.

## Suzhou

Sometimes called the "Venice of the East" for its many canals, Suzhou is best known for its beautiful classical gardens. The walled gardens were each created in a different style by Chinese scholars and artists between the 900's and 1900's. In the gardens, covered walks wind past ornamental ponds, summer houses, bamboo plants, and stones that are grouped to form tiny mountain ranges. The perfect harmony of the gardens expresses the Taoist ideal of achieving balance with nature and also reflects the quiet contemplation of Buddhism.

## Shanghai

Shanghai, China's largest city, lies on the Huangpu River about 14 miles (23 kilometers) from the Yangtze River. The city's location near these important waterways helps make it China's leading port and industrial city.

# BEIJING

Beijing, also spelled *Peking*, is China's capital and second largest city. Only Shanghai has more people. The city is part of the Beijing special municipal district, with a popultion of 11 million people. The district includes the central city, called the Old City, and a series of suburbs with farmland beyond.

Most of Beijing's residents belong to the Han ethnic group and speak Mandarin. Many of the minority groups in the city have adopted the customs and clothing of the Han.

Beijing, founded as a trading center nearly 4,000 years ago, has served as a center of government on and off for more than 2,000 years. *Beijing* means *northern capital*.

## The Old City

The Old City consists of two large rectangular areas called the Inner City and the Outer City. Walls once surrounded both areas, but they are gradually being torn down.

The Forbidden City and the Imperial City lie within the Inner City. The Forbidden City includes the palaces of former emperors. Only members of the emperor's household could enter the Forbidden City—hence its name. The buildings in this part of Beijing are now preserved as museums.

The Imperial City, which surrounds the Forbidden City, includes lakes, parks, and residences of China's Communist leaders. At the southern edge of the Imperial City stands the imposing *Gate of Heavenly Peace* (Tiananmen). This gate overlooks a huge square, the site of parades, fireworks displays, and many historic gatherings. On Oct. 1, 1949, Mao Zedong proclaimed the establishment of the People's Republic of China, with Beijing as its capital, to a crowd of thousands in Tiananmen Square. In June 1989, the square was the scene of another historic event when pro-democracy demonstrations led by Chinese students were brutally crushed by the nation's army.

The Beijing Zoo lies just outside the Inner City. Giant pandas, found only in western and southwestern China, are one of the zoo's main attractions.

The Temple of Heaven, or Tiantan, is a monument from China's imperial past. Architects from the Ming dynasty built the temple and many other buildings that still stand in Beijing's Old City.

Commercial areas, residential areas, and parks make up much of Beijing's Outer City. The Temple of Heaven, where Chinese emperors used to pray for a good harvest, stands at the southern end of the Outer City.

The Summer Palace, which served as an imperial residence during the summer, lies northwest of the Old City of Beijing. The tombs of Ming emperors and Beijing University are also located in the northwest suburbs.

Many people in the Old City live in old, one-story houses that border the *hutongs* (narrow, treelined alleys) that branch out from the main streets. Vendors walk up and down the hutongs selling such foods as fish, noodle soup, and vegetables.

Beijing's central business district includes the China's Central Television headquarters in the foreground and the China World Trade Center Tower 3 at the right. The tower is Beijing's tallest building.

Shaded by a large umbrella, a policeman directs traffic in Beijing's Tiananmen Square. As in other Chinese cities, bicycles make up most of the traffic. The Great Hall of the People stands in the background.

## Modern Beijing

After the Communists took over in 1949, China's new rulers modernized the city. By the end of the 1950's, several new buildings had been erected around Tiananmen Square. These included the Great Hall of the People (China's parliament building) and the National Museum of China. The city's population grew, and the economy expanded. City planners built many apartment buildings, Beijing's main railroad station, an underground commuter line, and a new traffic system.

Today, high-technology industries and tourism play a major role in Beijing's economy. Other industries include finance, banking, insurance, construction, and foreign trade.

Beijing was the site of the 2008 Summer Olympic Games. The city undertook a number of major construction projects to improve its services and facilities for the games. These included the National Center for the Performing Arts, a massive performance hall; Terminal 3 at Beijing airport, the largest building in the world; and Beijing National Stadium, nicknamed "the Bird's Nest," which hosted the Olympics' opening and closing ceremonies, as well as many of its competitions.

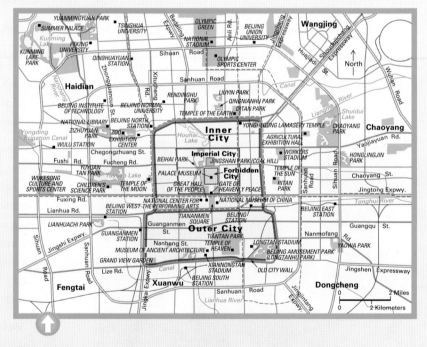

Beijing's Old City consists of the Inner City and the Outer City. The Inner City contains the Forbidden City, the palace of Chinese emperors, as well as the Great Hall of the People fronting on Tiananmen Square. The Great Hall of the People is the home of the National People's Congress and many government offices. East and south of the Old City lie suburbs, where most of Beijing's factories are located. Part of the Great Wall of China runs just north of Beijing.

# HONG KONG

Hong Kong is a special administrative region of China. It lies on the southern coast of China, near the mouth of the Zhu Jiang (Pearl River). It is southeast of China's Guangzhou Province.

Hong Kong consists of a peninsula, which is attached to mainland China, and more than 260 islands. The mainland area has two sections—the New Territories in the north and the Kowloon peninsula in the south. The main island, Hong Kong Island, lies south of the peninsula. Hong Kong covers a land area of only about 426 square miles (1,104 square kilometers), but about 6 million people live there. Hong Kong is one of the world's most densely populated places, with more than 15,500 people per square mile (6,000 per square kilometer).

Hong Kong, once a British colony, is a special administrative region on China's southern coast.

Rugged mountains and rolling hills cover much of Hong Kong. Some mountains in the New Territories rise more than 3,000 feet (910 meters) above sea level. Tai Ping Shan, on Hong Kong Island, is 1,818 feet (554 meters) high.

Only about 8 percent of Hong Kong's land is suitable for agriculture. Poultry farms and vegetable and flower fields are scattered throughout the New Territories, but most people who live in rural villages work in services or industries in the urban areas. The region depends heavily on imports for its food.

Almost all of Hong Kong's people live in urban areas. The entire northern coast of Hong Kong Island, the southern tip of the Kowloon Peninsula, and the satellite cities in the New Territories form a major metropolitan area. The metropolis once consisted of individual cities called Victoria (on the northern part of Hong Kong Island) and Kowloon (on the southern Kowloon Peninsula). The old city boundaries are rarely recognized today. Victoria, now known simply as Central, is Hong Kong's seat of government

Victoria Harbour separates Hong Kong Island from the Kowloon Peninsula. Motor-vehicle tunnels run underneath the harbor.

Chinese and Western advertising dominates a wide shopping street in Hong Kong. Because there are few import duties, many products can be bought and sold more cheaply in Hong Kong than in most other parts of the world.

## An economic center

Hong Kong is a center of international trade, finance, and tourism. Most workers are employed in service industries in these fields. Hong Kong's economy is highly integrated with that of China, and many international business firms maintain offices in Hong Kong from which they carry on business with China.

Hong Kong is a *free port*—that is, it collects no import duties on goods brought in from elsewhere, except for such goods as alcohol, tobacco, and perfume. Because there are few import duties, many products can be bought and sold more cheaply in Hong Kong than in most other parts of the world.

## Hong Kong's history

Hong Kong was part of China from ancient times until the 1800's. Then, through treaty agreements, Hong Kong became a British dependency. Hong Kong came under Chinese control again on July 1, 1997.

About 95 percent of Hong Kong's nearly 6 million people are Chinese. The non-Chinese residents include people from Australia, Canada, Indonesia, the Philippines, the United Kingdom, and the United States.

British control of Hong Kong began when the United Kingdom acquired Hong Kong Island in 1842 and the Kowloon Peninsula in 1860 through treaty agreements with China. In 1898, China leased the New Territories and some smaller islands to the United Kingdom for 99 years.

In 1842, Hong Kong had only about 5,000 people, but its population increased dramatically during the 1900's, when several waves of immigrants flooded the region. Many Chinese fled to Hong Kong during periods of upheaval in China. For example, Hong Kong's population grew to 500,000 after revolutionaries overthrew China's Manchu dynasty and established the Republic of China in 1912.

After the Communists took over China in 1949, Chinese immigrants pushed the number of people in Hong Kong up to about 2 million. Then, in the late 1970's and the 1980's, thousands of Vietnamese fled Communist rule in their country and migrated to Hong Kong.

The Chinese Communist government never formally recognized the United Kingdom's control of Hong Kong, but it did not oppose it either, because Hong Kong was of great value to China's economy. China earned money by selling food, water, raw materials, and manufactured products to Hong Kong. The Chinese government also owned a great deal of property in the region.

Nonetheless, in 1984, China and the United Kingdom agreed to transfer Hong Kong from British to Chinese rule in 1997. Under the agreement, Hong Kong became a special administrative region (SAR) of China with a high degree of control over its affairs, except in foreign policy and defense. The Hong Kong SAR retains its own executive, legislative, and judicial power. It is also allowed to preserve its capitalistic economy. These arrangements are to be in effect for at least 50 years after 1997.

After the signing of the agreement, economic cooperation between Hong Kong and China increased. Hong Kong industrialists continued the trend, begun in the 1980's, of moving manufacturing activities to China to take advantage of inexpensive labor available there. This trend continued into the early 2000's. Today, most employment in Hong Kong has shifted from manufacturing to service industries.

# MACAU

Macau, a special administrative region of China, lies at the mouth of the Zhu Jiang (Pearl River), about 40 miles (64 kilometers) west of Hong Kong. The territory has an area of only about 8 square miles (21 square kilometers) and consists of the city of Macau, which occupies a peninsula, and the small islands of Taipa and Coloane. More than 95 percent of the territory's population of about 435,000 people is Chinese. A small fraction of the population is of Portuguese ancestry.

A busy cook prepares Chinese and Portuguese snacks at a roadside stand in Macau's "night market."

A Chinese astrologer in Macau offers her predictions to customers at an outdoor stand.

Some sections of Macau include modern high-rise hotels, apartments, and office buildings. Gambling casinos, racetracks, and night clubs attract many tourists, mainly from Hong Kong. Macau's economy is based on tourism and light industry, chiefly the manufacture of fireworks, furniture, textiles, and toys.

## A blend of cultures

Macau combines both Chinese and Portuguese influences. Macau was formerly a territory of Portugal, and the Portuguese presence there dates back to the

1500's. In some areas of Macau, cobblestone streets are lined with old, pastel-colored houses much like those in Lisbon, the capital of Portugal.

Roman Catholic churches in Macau resemble those built between the 1400's and the 1600's during Portugal's golden age of art. The facade of the São Paulo Cathedral, which is all that remains of the church after a fire in 1835, is one of the territory's most famous sights. Roman Catholic churches, however, are often located near Buddhist shrines, such as the ancient temple dedicated to A-Ma, the goddess of sailors and fishermen. Christian Portuguese settlers named Macau after this goddess. In addition, both Catholic and Buddhist festivals are celebrated in Macau.

Even the different cuisines of the two regions have been combined. Restaurants serve Chinese specialties with Portuguese wine, and traditional Portuguese fish stew is eaten with chopsticks.

The ruined São Paulo Cathedral is one of Macau's most famous landmarks. Only the cathedral's ornate facade survived the fire that destroyed the rest of the building in 1835.

Macau lies at the mouth of the Zhu Jiang (Pearl River) on China's southern coast.

Macau's past as a Portuguese colony is reflected in its pastel-colored buildings that recall Lisbon. The pedicabs parked in front of the buildings are, however, typical of the Far East.

Macau's casinos attract many visitors to the "Las Vegas of the East."

## Past, present, and future

The Portuguese were the first Europeans to establish trade with China. In 1557, they leased Macau from China and established a Portuguese settlement in the territory. Macau soon became the chief port for Portugal's trade operations in China and Japan. It also served as the headquarters for Catholic missions in East Asia. A Portuguese governor was appointed to rule Macau's Portuguese settlers in 1680, but the Chinese retained complete authority over the territory itself and its native population.

Although Portugal's power had begun to decline by the 1600's, it held on to much of its empire—including Macau—well into the 1900's. During China's Cultural Revolution of the late 1960's, Portugal offered to return Macau to the Chinese, but China's Communist rulers refused the offer.

In the 1970's, Portugal's government ended the country's control of its colonies, but Macau remained a Portuguese territory. China allowed Portugal to retain Macau because Macau was contributing to China's economy. Macau bought almost all its food and drink-

ing water from China, and these purchases provided China with foreign currency for international trade.

Portugal returned Macau to China in December 1999 under an agreement signed by both countries in 1987. As a special administrative region of China, Macau maintains separate political, judicial, and social systems from the rest of the country. China, however, appoints the chief executive of Macau.

# THE YANGTZE RIVER

The Yangtze River runs for about 3,915 miles (6,300 kilometers) across China. The Yangtze is the third longest river in the world, and the longest and most important river in China. To most Chinese, the Yangtze is known as the *Chang Jiang*, or *long river*.

The Yangtze River has been called "China's equator" because it divides the wheat-growing northern half of the country from the rice-growing south. A large percentage of China's rice is grown along the Yangtze, as well as much of the country's cotton, corn, and wheat.

This great river has been one of China's most important transportation waterways for hundreds of years. In the 1200's, Marco Polo marveled at the number of ships on the Yangtze. Ships and boats of all sizes still carry cargo on the Yangtze today. About half of China's ocean trade is distributed over the Yangtze and its branches.

Ocean steamers reach Wuhan—680 miles (1,090 kilometers) by river from the coast. Cargo boats and small wooden sailing craft called *junks* also travel the river, and thousands of Chinese live on the Yangtze in junks.

The Yangtze River begins in the Tanggula Mountains of Qinghai Province near Tibet at an altitude of about 16,000 feet (4,880 meters). After crossing much of China, its waters empty into the East China Sea, just north of Shanghai.

The Yangtze begins as little more than a trickle of water, flowing across the Plateau of Tibet in a shallow, wide valley. The high mountains at the Yangtze's source cause it to

At 3,915 miles (6,300 kilometers) in length, the Yangtze is the longest river in China and the third longest in the world. It divides the wheat-growing northern half of China from the rice-growing south.

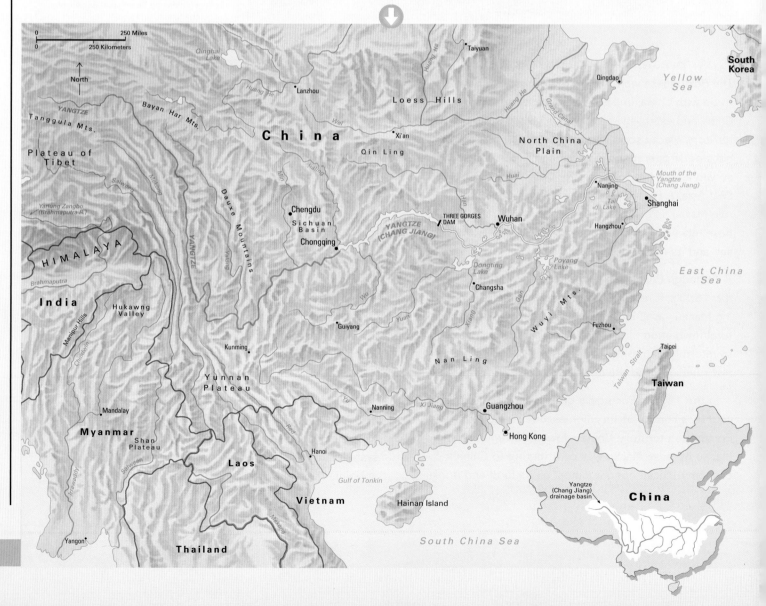

The rapid drop of the Yangtze as the river goes south from the plateau of Yunnan increases the speed at which the water moves. Deep gorges have been cut by the fast-flowing water.

flow rapidly for most of its length. When the river goes south from the plateau to Yunnan, the rapid drop in elevation increases the speed at which the water moves. Deep gorges have been cut by the fast-flowing river in its upper course above Yichang.

From Yunnan, the Yangtze flows northeast across the Sichuan Basin. Here, too, the river has carved great gorges that make land travel difficult. Ships can travel east on the Yangtze into western Sichuan, but only small craft can navigate the four swift-flowing tributaries that flow into the river in this region. In Sichuan, the Yangtze passes through Chongqing, an inland port and major center of industrial activity.

The Three Gorges Dam near Yichang is the world's largest dam. Designed to control flooding, improve navigation, and generate hydroelectric power, it has also caused controversy, because the dam and reservoir will displace more than a million people. Construction began in 1994. It was completed in 2010.

The Han River flows into the Yangtze at the industrial area of Wuhan. The Yangtze is almost 1 mile (1.6 kilometers) wide at this point, and many wharves and bridges line its banks. From Wuhan, the river flows across a broad plain, one of China's most heavily developed industrial and agricultural regions. Occasional floods in this area have driven many people from their homes and severely damaged crops and property.

The Yangtze then moves northeast to Nanjing and enters the Yangtze River Delta, which has the best combination of level land, fertile soil, and sufficient rainfall in all China. The so-called Fertile Triangle between Nanjing, Shanghai, and Hangzhou is one of the most densely populated rural areas in the world.

At Hangzhou, the Grand Canal, a water-way completed around A.D. 610, links the Yangtze with the Yellow River and China's northern cities. About 14 miles (23 kilometers) north of Shanghai, the Yangtze and the Huangpu rivers meet and empty into the East China Sea.

# AGRICULTURE

Agriculture is the foundation of China's economy. About 55 percent of the people live in rural villages, and about half of all workers are farmers. In southeastern and east-central China, farmers grow cotton, rice, sweet potatoes, tea, and wheat. Corn and wheat are important crops in the northeast. China produces more apples, cabbages, carrots, cotton, pears, potatoes, rice, sweet potatoes, tobacco, tomatoes, and wheat than any other country. It is also a leading producer of corn, melons, rubber, soybeans, sugar beets, sugar cane, and tea. Farmers in the far south grow tropical crops, such as bananas, oranges, and pineapples.

Many rural families raise chickens and ducks. China has more domesticated ducks than any other country. China also has nearly half the hogs in the world. Farmers raise them for meat and fertilizer. Farmers also raise cattle, goats, horses, and sheep.

## Bountiful harvests

Only about 15 percent of China's land is suitable for farming, and droughts and floods often interrupt production. Nevertheless, the country produces almost enough food for its huge population. The long growing season in southern China, where two or more crops can be grown on the same land each year, is an important factor in this achievement. Chinese farmers also practice *terracing*—growing crops on level strips of land cut out of hillsides—to utilize the fertile land in hilly or mountainous areas. In addition, they make extensive use of irrigation and organic fertilizers. They also practice soil conservation.

China's farm output has greatly expanded since the Communists took control of China. Although economic growth was slow at first, a series of reforms launched in 1978 vastly increased productivity.

Farmers sell their own produce at an open market. Under the current agriculture system, each farm family assumes full responsibility for the entire process of production—from selecting the seeds to gathering the harvest.

A bridge spans a stream between terraced rice fields in southern China. Farmers use every bit of land they have to support themselves and feed the huge population of China.

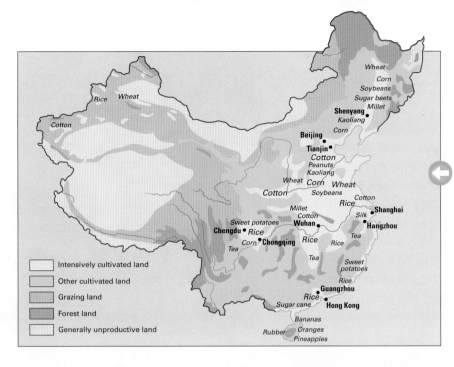

China's huge farm output makes it the world's largest producer of agricultural products. However, the country's average per capita (per person) income is still low, and many economists still consider China a developing country.

Map labels: Wheat, Corn, Soybeans, Sugar beets, Millet, Shenyang, Kaoliang, Corn, Beijing, Tianjin, Cotton, Peanuts, Kaoliang, Wheat, Corn, Wheat, Cotton, Soybeans, Cotton, Rice, Shanghai, Silk, Millet, Cotton, Hangzhou, Sweet potatoes, Wuhan, Tea, Chengdu, Rice, Corn, Chongqing, Rice, Rice, Tea, Tea, Sweet potatoes, Rice, Rice, Guangzhou, Rice, Hong Kong, Sugar cane, Bananas, Rubber, Oranges, Pineapples, Rice, Wheat, Cotton

Legend:
- Intensively cultivated land
- Other cultivated land
- Grazing land
- Forest land
- Generally unproductive land

## Reforms and counter-reforms

During the 1950's, the Communists seized control of China's farmland and organized the peasants to farm the land cooperatively in units called *communes*. However, the commune system was not a success. Between 1957 and 1978, grain production rose only slightly, and China had to import large quantities of grain to feed its huge and increasing population. In addition, the standard of living on farms had not improved, and Chinese farmers were discouraged.

In December 1978, the government decided that fundamental reforms in agriculture were needed and began loosening some of its control. Many months of discussions led to the adoption of a new policy that allowed individual families to farm more of the land. The government also worked to increase farm production by using higher-yielding seeds, more machinery, more irrigation, and more chemical fertilizers.

Today, cooperative groups known as *collectives* make contracts with individual families. The contract tells how much land a family can work, what crops and livestock the family will raise, and how much it will sell to the government at a set price. After fulfill-ing its contract, the farm family may use the rest of its production as it wishes. Most families use some for food and sell the rest on the open market.

Some rural families sign contracts as *specialized households*. They may specialize in raising only one commodity, such as chickens or silk. In some cases, they may provide farm machinery, repairs, or handicrafts on the free market instead of doing full-time farm work. After paying an agreed amount to the government, the household keeps any profit. Some households operate businesses or small factories and hire employees.

As a result of these reforms, agricultural productivity has risen. Most important, the standard of living in rural areas has greatly improved.

## Forestry and fishing

China is a leading producer of forest products. The timber industry is located mainly in northern China. China also has the world's largest fishing industry, with an annual catch of millions of tons of seafood. Fish farming is an important industry. Fish farmers raise fish in ponds both for food and for use in fertilizer.

# INDUSTRY

In the early 1950's, China's industrial structure was modeled after the system that was in use in the Soviet Union. The government owned and operated all public enterprises and emphasized the development of heavy industry. Today, China has one of the world's largest steel industries. The machine-building industry provides metalworking tools and other machines for new factories. Other heavy industrial products include cement, fertilizer and other chemicals, irrigation equipment, military equipment, ships, tractors, and trucks.

China has much of the energy and raw materials it needs to run these industries. The country is one of the world's largest producers and users of coal, and it also has rich deposits of oil. In addition, China is a leading producer of iron ore, magnesium, tin, and tungsten. China also mines aluminum, copper, gold, lead, silver, and zinc.

A employee works in a textile factory in Yunnan province. China is the world's leading producer of textiles.

## Limited growth

Although the Chinese government achieved some industrial growth under its hard-line Communist policies, the Communist system also hampered efficiency, productivity, and incentive. All the workers were paid the same wage—regardless of the quality of their work—and all factories were supported by the government regardless of their performance.

In 1978, to help China's workers reach their full potential, the government developed a plan for restructuring industries similar to the strategy they had used in the agricultural sector.

## Industrial reforms

Under the reforms, an industrial enterprise paid a quota of its profit to a supervisory board but kept a share of any profits over and above the quota. These profits could be used for bonuses, employee benefits, and industrial improvements. The government, led by moderate reformers, envisioned a mixed economy composed of both state-owned and privately owned enterprises. China also launched a more open foreign-trade program designed to help finance the modernization of its industries.

A worker arranges aluminum products in Dalian in Liaoning province. China outranks all other countries in the production of aluminum as well as lead, magnesium, tungsten, and zinc.

As industry rapidly progressed, the industrial workers' standard of living improved. People's demand for such consumer goods as television sets, refrigerators, and washing machines increased. Many of these items had to be imported because China's own consumer goods production was basically limited to textile manufacturing and food-processing.

By mid-1988, however, China was not making enough money on exports to pay for all the imported goods that consumers wanted. Too much money was chasing too few goods—a classic cause of inflation.

Moderate Communist Party leaders argued that bolder reforms and increased free enterprise would solve China's economic problems. Conservative leaders, on the other hand, wanted more cautious changes that would ensure continued state control of the economy.

In June 1989, the student demonstrations in Beijing brought the debate to a head. The conservatives felt that the students were challenging China's Communist leadership, and they saw a clear connection between the economic reforms and the protests. The military put down the demonstration by force, killing hundreds of demonstrators.

The national government still exercises much control over China's economy. It controls the most important industrial plants and operates most of the nation's banks, most long-distance transportation, and foreign trade. It also sets the prices of certain key goods and services.

Since the 1980's, however, the number of privately owned and operated businesses has increased dramatically. In addition, in 2001, China joined the World Trade Organization (WTO), which promotes international trade. Many experts considered China's membership in the WTO to be a sign of progress toward freeing the economy from government control.

China's booming economy since the late 1900's has raised living standards for most of its people. China has increased the production of consumer goods as demand from Chinese consumers has continued to grow. The largest consumer goods industries are textiles, food-processing, and electronics.

Chinese industry is heavily concentrated in the eastern half of the country. Shanghai is one of the world's leading manufacturing centers. Its output far exceeds that of any other city in China. Beijing, Tianjin, and Guangzhou are also important centers of industry.

Workers assemble an automobile in a factory in Hefei in Anhui province. The rising standard of living in China has led to an increased demand for autombiles and other consumer goods.

# TIBET

Tibet, a land in south-central Asia, is often called the *Roof of the World*. Its snow-covered mountains and its vast, windswept plateau are the highest in the world. Even the valleys in Tibet are higher than the mountains of most other countries. Some of Asia's greatest rivers begin in the Tibetan mountains, including the Brahmaputra, Indus, Mekong, Salween, and Yangtze rivers.

Although Tibet has been a part of China since the 1950's, it was an independent or semi-independent state for many years. While Tibet carried on some trade with other lands, its mountain ranges generally isolated the area from the outside world. Tibetans have sometimes been called the *hermit people*.

Tibetans are intensely religious people who turn prayer wheels and recite prayers on the streets. Tibet's religion is a branch of Buddhism called *Lamaism,* and the region was traditionally a *theocracy* (religious kingdom). Before China took control, Buddhist monks had a strong voice in the rule of Tibet.

The Tibet Autonomous Region of China has an area of 471,700 square miles (1,221,600 square kilometers) and a population of about 2.5 million. Prior to the Chinese take-over, Tibet covered about 965,000 square miles (2,500,000 square kilometers). The high, cold Plateau of Tibet covers most of the land. About 6 million Tibetans live throughout the plateau.

High mountains border the Plateau of Tibet, including the snowy Himalaya along its southern end. The Himalaya, which rises higher than any other mountain chain in the world, includes Mount Everest, the world's highest mountain at 29,035 feet (8,850 meters) above sea level. Ger, located more than 15,000 feet (4,570 meters) above sea level in western Tibet, is believed to be the highest town in the world.

The Plateau of Tibet is dotted with hundreds of salt lakes and marshes, but many of them have a high salt content and barren shores. Tibet also has hilly grasslands and forests in its northern sections. However, large parts of the region are wastelands of gravel, rock, and sand, where only the hardy Tibetan reed grass grows.

Green fields line a fertile valley in southern Tibet. Most Tibetans live in the south, where the climate is less harsh and the soil is more suitable for farming.

Tibet, part of China since the 1950's, lies in south-central Asia. It is bordered by India, Nepal, Bhutan, and Myanmar on the south and west.

Domestic yaks can carry heavy loads for 20 miles (32 kilometers) a day. The yak—well-suited for traveling across Tibet's rugged terrain—can slide down icy slopes, swim swift rivers, and cross steep rockslides.

The ruins of the Gandan monastery perch on the Plateau of Tibet. Many Lamaists left Tibet after the Chinese invaded in 1950 and removed the Buddhists from power.

The snow-covered peaks of the Himalaya enclose the sparkling waters of a mountain lake in southern Tibet. Many peaks of the Kunlun range in northern Tibet rise more than 20,000 feet (6,000 meters).

Because of the poor soil and cold climate, most of the land cannot be farmed, but fertile areas in southern Tibet allow some farming and livestock raising. In the shadow of towering mountain peaks, farmers grow their crops—chiefly barley—and raise cows, goats, horses, poultry, and sheep. Most Tibetans live in the sheltered valleys of the south and southeast. Lhasa, Tibet's capital and largest city, stands in this region.

Nomads roam the northern grasslands with their sheep and yaks. The domesticated yak is used as a beast of burden in Tibet. The yak also supplies butter, cheese, meat, and milk, as well as hides for tents and shoes. Tibet's wild creatures include deer, gazelles, tigers, bears, monkeys, pandas, and wild horses. Some of these animals feed on the thousands of different kinds of plants in Tibet. Dense forests of oak, pine, and bamboo flourish in the east and southeast.

Tibet has a severe climate. Winter brings sudden blizzards and snowstorms that can last for days. Fierce windstorms sweep Tibet in all seasons. January temperatures average 24° F (–4° C), but often drop as low as –40° F (–40° C). July temperatures average 58° F (14° C). In addition, because the towering Himalaya shuts out moisture-bearing winds from India, much of Tibet receives less than 10 inches (25 centimeters) of rain annually.

# TIBETAN PEOPLE AND HISTORY

Most of Tibet's people live in southern Tibet, and many work as farmers. In Lhasa, which ranks as the region's largest city with a population of about 140,000, many people hold jobs in government, light industry, or tourism.

Although Mandarin Chinese is the official language, Tibet's main traditional language is Tibetan. Both languages are taught in schools, and all government documents are written in both languages.

Traditional Tibetan houses are small, one- or two-floor homes with stone or brick walls and flat roofs. Animals are housed on the ground floor. Many Tibetans work in their homes in such traditional household industries as cloth weaving and carpet making. Wool is a major export.

Much of Tibetan life revolves around Lamaism. Religion is an important part of life, and festivals in Tibet are religious in character. Many people make long pilgrimages to important temples in Lhasa or Xigaze. The traditional emphasis on religious life has decreased, however, as a result of Communist Chinese rule.

## Early history

During the A. D. 600's, Tibet became a powerful kingdom that played an important role in central Asia for 200 years. During this time, Buddhism and writing were introduced from India, and Lhasa was founded. Buddhism was eventually combined with traditional Tibetan religious beliefs to form Lamaism. The Lamaists built monasteries called *lamaseries,* which became political and educational centers in Tibet.

Between 900 and 1400, several Lamaist sects developed in Tibet. The most powerful was the Yellow Hat sect—so called because its monks wore yellow uniforms. Their

A carpet seller, warmly bundled against the cold Tibetan winds, displays his wares on a street in Lhasa. Carpets are woven from yak wool, which is also used to make winter clothing. Lighter garments are made of hemp and cotton.

A rural trader fills a bucket with rich yak milk. In Tibet, yaks serve many of the same purposes served by cows in Western countries. The domestic yak even provides butter, which, together with salt and soda, is used to flavor tea.

The Potala Palace in Lhasa has gold-leaf roofs and more than 1,000 rooms. Formerly a residence of the Dalai Lama and other monks, it now is a museum that houses many art treasures.

China seized control of the media, banks, and food stores. The Chinese also took most of the best jobs, with many working as government administrators and teachers. In 1959, Tibet rebelled against Chinese rule. About 87,000 Tibetans were killed in the uprising, and the Dalai Lama fled to exile in India.

The people of Tibet continued to stage riots against the Chinese in the 1960's. The Chinese retaliated by closing or destroying most of Tibet's monasteries. In the 1980's, the Chinese government adopted a more liberal policy toward Tibet. Some monasteries were reopened, and farmers were allowed to decide for themselves which crops they would grow and sell. But Tibetan demonstrations against Chinese rule halted these reforms.

In 1989, the Panchen Lama died. The Panchen Lama and Dalai Lama are the two main spiritual leaders of Tibetan Buddhism. In 1995, the Dalai Lama announced the selection of a new Panchen Lama, but the Chinese government installed a rival candidate of its own.

In 2008, protests calling for Tibetan independence turned into riots. The Chinese government sent troops to restore order. In 2011, the Dalai Lama gave up his political leadership, though he remained a spiritual leader. Members of the international Tibetan community elected a new prime minister of the Tibetan government in exile.

leader, the Dalai (High) Lama, became the spiritual and political leader of Tibet in 1642.

China controlled Tibet from the early 1700's until 1911, when Tibetans forced out the Chinese troops. However, in 1950, China—under Communist rule—once again invaded Tibet. The Chinese claimed that they wished to liberate Tibetan farmers from serving the nobility and monks who owned the farmland. Most farmers were *serfs*—workers who were not free to leave the land—and much of what they produced had to be given to the landowners. In 1951, Tibet surrendered its sovereignty to the Chinese government, and China broke up the large estates of the monks and the nobility and distributed them among farmers.

## Communist Chinese rule

The Communists agreed to allow Tibet regional self-government. They also promised to guarantee the Tibetans freedom of religious belief. However, by 1956, China had begun tightening its control of the region. The Chinese army forced the Tibetan peasants, who were used to growing barley, to grow wheat to feed the Chinese soldiers.

Prayer flags and religious paintings decorate the rocks near a Tibetan monastery. Many of Tibet's monasteries were destroyed by the Chinese in the 1960's and 1970's, and many monks now work in agriculture and handicrafts.

The fourth largest South American country in area, Colombia is situated on the extreme northwest part of the continent. It is a nation with a troubled history. Years of suffering and devastation under Spanish rule were followed by periods of political disorder, as the country struggled to establish an orderly society and a democracy.

In the late 1900's, the illegal activities of the country's infamous drug dealers caused a serious breakdown in law and order for Colombia. In addition, natural disaster struck Colombia in 1985, when the Nevado del Ruiz volcano erupted twice, killing about 23,000 people.

Colombia boasts some of the most varied and magnificent scenery in all of South America. Miles of sandy beaches lined with graceful palm trees stretch along its Caribbean and Pacific coasts, while the snow-covered peaks of the Andes tower high above the landscape. Banana, cotton, and sugar cane plantations dot the sun-drenched lowland plains. In the south, tropical rain forests rise up along the rivers that feed into the great Amazon.

## A rich heritage

Adding to the charm of the Colombian countryside is the rich cultural heritage of its people. In the southern Andean region, for example, gigantic earthen mounds, stone statues, and tombs around San Agustín date from about the first 900 years A.D.—the relics of an advanced civilization.

Today, many exquisite works created hundreds of years ago by *indigenous* (native) goldsmiths are preserved in the famed *Museo del Oro* (Gold Museum) in Bogotá, Colombia's capital city. These masterpieces, including small statues and pieces of jewelry, are ranked among Latin America's greatest artistic treasures.

The most famous indigenous goldsmiths were the Chibcha Indians, who made *tunjos*—flat, stylized human figures with long bodies, large heads, and decorations of gold wire. The Chibcha were farmers who lived in the eastern chain of the Andes, near the center of what is now Colombia. During Chibcha coronation ceremonies, the new ruler was covered with gold dust to represent a god. Then he was rowed out to the middle of a lake and washed free of the gold dust to represent a human ruler.

# COLOMBIA

The legend of El Dorado—a kingdom filled with gold and riches—grew out of this custom. Under the leadership of Gonzalo Jiménez de Quesada, Spanish explorers in search of El Dorado mercilessly conquered the Chibcha and founded the city of Bogotá in 1538. Jiménez de Quesada called the surrounding area the New Kingdom of Granada because it reminded him of the region in Spain known as Granada.

The Spaniards gradually spread throughout the region. The conquerors used the indigenous peoples as slave laborers, forcing them to mine emeralds, gold, and platinum. They also brought Africans to work as slaves on sugar cane and cacao plantations along the Caribbean coast.

## Independence

Colombians celebrate July 20, 1810, when Bogotá established a government independent of Spain, as their independence day. Most Spanish provinces in South America also set up independent governments at that time. Spain tried to regain control but was unsuccessful. After bitter fighting, Colombia broke completely free from Spanish rule in 1819, when the Venezuelan general Simón Bolívar defeated Spain in the Battle of Boyacá, near Bogotá. The republic of Gran Colombia was established, which included what are now Colombia, Ecuador, Venezuela, and Panama. By 1830, Ecuador and Venezuela had become separate nations. Colombia lost Panama in 1903.

Almost from the beginning, the new nation was torn by disputes over how strong the central government should be. Those who favored a strong central government and a powerful role for the Roman Catholic Church became the Conservative Party. Members of the Liberal Party preferred strong regional governments.

Over the years, fighting between the groups often led to civil unrest and rioting, reaching a climax between 1948 and the 1960's. About 200,000 Colombians died during that period, known as *La Violencia* (The Violence).

# COLOMBIA TODAY

Unlike most other Latin American countries, Colombia has had a government elected by the people through much of its history. Colombia is a republic, headed by a president who is elected to a four-year term. Voters also elect the members of the two houses of Congress to four-year terms.

## Liberals and Conservatives

Despite its long tradition of democracy, Colombia has suffered years of political disorder and civil strife. By 1957, the turmoil created by *La Violencia* forced the Liberal and Conservative parties to form a *coalition* (joint government) to restore order in the country. Between 1958 and 1974, they shared all political offices, and the leaders of each party alternated as the nation's president every four years. The coalition, known as the National Front, restored the people's confidence in their government and improved the economy.

The two parties have come to share many beliefs since the formation of the coalition. Nevertheless, their traditional differences remain. The Liberal Party supports regional government, religious tolerance, and the social and economic demands of the masses. The Conservative Party prefers a centralized government managed by a small elite.

## A history of violence

Violence erupted again in the mid-1960's, when left-wing guerrilla groups began fighting against the government. The two largest leftwing groups are the Revolutionary Armed Forces of Colombia (FARC) and the National Liberation Army (ELN). Each group supports establishing its own version of Communist government, through armed revolution.

In the 1970's, several criminal groups in Colombia began to make huge profits from producing and selling illegal drugs, especially cocaine. Colombia is the world's largest producer of cocaine. The United States has been the largest market for the drugs. In the 1980's and early 1990's, the drug trade was controlled by *cartels* (associations formed by suppliers to control the market for their product). The largest cartels were based in Medellín and Cali.

## FACTS

| | |
|---|---|
| Official name: | Republica de Colombia (Republic of Colombia) |
| Capital: | Bogotá |
| Terrain: | Flat coastal lowlands, central highlands, high Andes Mountains, eastern lowland plains |
| Area: | 440,831 mi² (1,141,748 km²) |
| Climate: | Tropical along coast and eastern plains; cooler in highlands |
| Main rivers: | Cauca, Magdalena, Guaviare, Meta, Caquetá |
| Highest elevation: | Cristóbal Cólon, 18,947 ft (5,775 m) |
| Lowest elevation: | Pacific Ocean, sea level |
| Form of government: | Republic |
| Head of state: | President |
| Head of government: | President |
| Administrative areas: | 32 departamentos (departments), I distrito capital (capital district) |
| Legislature: | Congreso (Congress) consisting of the Senado (Senate) with 102 members serving four-year terms and Camara de Representantes (House of Representatives) with 166 members serving four-year terms |
| Court system: | Corte Suprema de Justicial (Supreme Court of Justice), Council of State, Constitutional Court |
| Armed forces: | 267,200 troops |
| National holiday: | Independence Day - July 20 (1810) |
| Estimated 2010 population: | 46,271,000 |
| Population density: | 105 persons per mi² (41 per km²) |
| Population distribution: | 73% urban, 27% rural |
| Life expectancy in years: | Male, 69; female, 76 |
| Doctors per 1,000 people: | 1.4 |
| Birth rate per 1,000: | 20 |
| Death rate per 1,000: | 6 |
| Infant mortality: | 19 deaths per 1,000 live births |
| Age structure: | 0-14: 30%; 15-64: 65%; 65 and over: 5% |
| Internet users per 100 people: | 38 |
| Internet code: | .co |
| Languages spoken: | Spanish |
| Religions: | Roman Catholic 90%, other 10% |
| Currency: | Colombian peso |
| Gross domestic product (GDP) in 2008: | $244.24 billion U.S. |
| Real annual growth rate (2008): | 3.5% |
| GDP per capita (2008): | $5,663 U.S. |
| Goods exported: | Bananas, coal, coffee, copper, flowers, petroleum, textiles |
| Goods imported: | Chemicals, machinery, transportation equipment |
| Trading partners: | Brazil, China, Germany, Japan, United States, Venezuela |

High-rise office and apartment buildings are changing the skyline of Colombia's cities. The Spanish-style homes built during the colonial era—adobe structures with red tile roofs and patios—are rapidly disappearing.

Political violence and the drug trade have become intertwined. In the 1980's, the Colombian and U.S. governments began a joint effort to stop the drug trade. In response, the cartels mounted a campaign of bombings, assassinations, and other terrorist acts. At times, the drug cartels have arranged alliances or pay-offs with both right-wing paramilitary groups and left-wing rebel groups in return for support and protection.

In the late 1980's, the government reached an agreement with some of the guerrilla groups. The peace process led to the creation of a constituent assembly that adopted a new constitution in 1991. The FARC and ELN, however, did not up-hold the peace process.

During the 1990's, several right-wing paramilitary groups banded together as the United Self-Defense Forces of Colombia (AUC) to fight leftist guerrillas. The paramilitaries, which received money from drug trafficking, committed murders, kid-nappings, and other crimes against both guerrilla soldiers and civilians. In 2004, the Colombian government conducted negotiations that arranged for many AUC members to disarm, but many people doubted they would remain unarmed.

In 1998, the government began peace talks with the left-wing FARC and ELN. The government withdrew its forces from about 16,000 square miles (41,400 square kilometers) of territory in southern Colombia, giving the FARC control over the area. But when the FARC and ELN continued to enlarge their forces and launch terrorist attacks, the government forces returned in 2002.

Àlvaro Uribe Vélez was elected president of Colombia in 2002. Uribe took a strong stand against left-wing drug traf-fickers. Colombia's economy grew during his administration, and he was reelected in 2006. In 2010, Juan Manuel Santos, who was widely considered Uribe's polit-ical heir, was elected president.

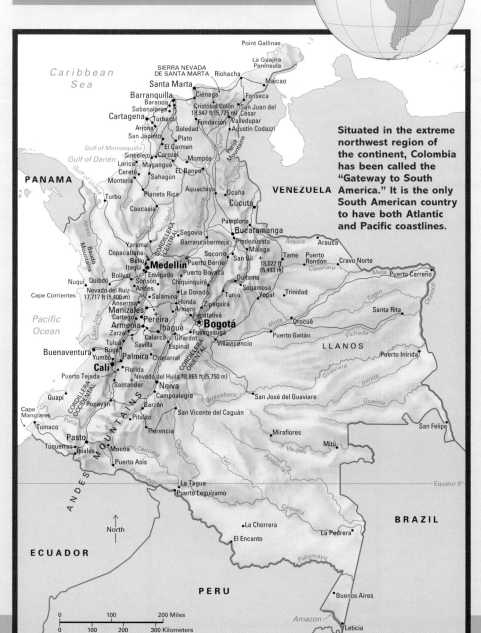

Situated in the extreme northwest region of the continent, Colombia has been called the "Gateway to South America." It is the only South American country to have both Atlantic and Pacific coastlines.

# LAND AND ECONOMY

Colombia's great variety in climate and landscape make it a land of seemingly endless contrasts. Although the country is situated entirely within the tropics—the equator crosses its southern region—parts of Colombia are quite chilly because of their high altitude. From the snow-capped peaks of the Andes Mountains in the west to the tropical rain forests in the south, Colombia's scenic landscape makes it a fascinating country to explore.

## Landscape and climate

The Coastal Lowlands, the Andes Mountains, and the Eastern Plains make up Colombia's three main land regions. Most of the country's people live in the valleys and basins of the Andes. Many of Colombia's jungles are so remote that they have yet to be explored and mapped in detail.

The Coastal Lowlands lie along the Caribbean Sea and the Pacific Ocean. The two coasts differ greatly in character. The Caribbean Lowlands have about 20 percent of Colombia's people and the nation's busiest seaports. The bustling cities of Barranquilla, Cartagena, and Santa Marta handle most of Colombia's foreign trade. Inland, large banana, cotton, and sugar cane plantations, as well as cattle ranches and small farms, dot the landscape. By contrast, the Pacific Lowlands consist mostly of swamps and dense forests where heavy rains fall almost every day.

Three ranges of the Andes Mountains, separated by Colombia's two major rivers, cover about a third of the country. The Magdalena River separates the middle and eastern ranges of the Andes, while the Cauca River flows between the middle and western ranges. About 75 percent of Colombia's people live within these mountain valleys and river basins. The region's rich mines, fertile farms, and large factories produce most of the country's wealth.

About 60 percent of Colombia's land is covered by the warm, humid Eastern Plains—a sparsely populated area. Many rivers flow eastward across these plains to join the Orinoco and Amazon systems. In the north, cattle and sheep graze on the rolling grasslands known as the *llanos,* while the southern forests are home to scattered Indian tribes.

Blue waters and sunny beaches make Colombia's Caribbean coast a beautiful tourist spot. Both government programs and private enterprises have developed resort hotels and other tourist facilities in Colombia.

Swift-flowing rivers cut deep valleys through the soaring Andes peaks where pure mountain waters tumble over the massive rocks. Farmers grow a wide variety of crops in the rich soil of the river valleys.

 A coffee farmer removes the husks from beans to prepare them for market. Coffee shrubs thrive in the mild climate of the steep Andes slopes. Colombia is one of the world's top coffee-producing countries. It provides about 10 percent of the world's trade in coffee.

## Economy

Colombia is a developing country. Its economy has long been dependent on agriculture, with coffee the leading legal export crop. Other major crops include bananas, *cassava* (a tropical plant with starchy roots), corn, cotton, flowers, potatoes, rice, and sugar cane. Agriculture employs about a third of all workers. Growing *coca,* the plant from which cocaine is made, is illegal but still continues.

Government leaders have worked to reduce the country's dependence on agriculture. Since the 1950's, manufacturing has grown steadily. Today, manufacturing employs about a fourth of Colombia's workers. Major manufactured products include cement, chemicals, metal products, processed foods and beverages, and textiles and clothing.

Colombia is rich in minerals, and mining is rapidly growing in importance. Colombia is one of the world's leading producers of emeralds. It also has gold mines.

Petroleum and coal are major exports. The country also has reserves of natural gas and iron ore. Large underground salt deposits provide raw material for the country's chemical industry.

A Colombian national policeman pulls up a coca plant in La Macarena National Park. The plant is used to make the illegal drug cocaine. The destruction of the coca plants is part of a government antidrug campaign, which is largely funded by the United States.

# PEOPLE

Like the people of most other South American countries, Colombians have widely different ethnic origins. A large *indigenous* (native American Indian) population lived in what is now Colombia when the first Spanish colonists arrived in the 1500's. After the Spaniards conquered the territory, they brought Africans to work alongside the indigenous people as slave laborers.

## Ethnic groups

Over the years, many indigenous, Spanish, and African people intermarried. *Mestizos*—people of mixed European and indigenous ancestry—make up most of Colombia's population. About 4 percent of Colombia's people are of unmixed African ancestry. Afro-Colombians (people of mixed African and indigenous ancestry) make up another 3 percent. About 80 indigenous groups make up about 1 percent of the population.

Each group has its own cultural heritage, but together these ethnic groups weave the rich tapestry that makes up the Colombian national character. For example, the lively Afro-Caribbean rhythms of *La Cumbia,* the national dance, are a fascinating contrast to the haunting melodies played on flutes, stringed instruments, and drums by the native people of the southern Andes.

Almost all Colombians speak Spanish, the country's official language. Colombians are proud of their language. They consider it closer to the pure Castilian Spanish of Europe than the Spanish spoken in other Latin-American countries. They have even passed a law to protect their language from unnecessary change.

## Way of life

About a quarter of Columbia's people live in rural areas. Rural families in the warm, wet coastal regions use bamboo poles and palm leaves to build well-ventilated houses. In the cooler mountain zones, many houses have thick adobe walls.

A small number of wealthy landowners own most of the farmland that is not used for coffee growing.

The central square in Silvia, a large town south of Cali in the Andes, provides a convenient spot for the residents to chat. Their traditional skirtlike garments and felt hats are now seen only rarely among people in Colombia's rural areas.

An indigenous Columbian paddles a canoe along the Putumayo River, which forms the southern border with Peru. Only a few native tribes remain in Colombia, mostly in remote areas of the country.

They hire workers or rent their land to tenant farmers. Most tenant farmers and small landowners work very small plots of land that are barely large enough to support their families.

A few small indigenous tribes inhabit the great forests bordering the Amazon Basin, where they fish, hunt animals for food, and gather wild rubber. Those who live in larger settlements on the dry La Guajira Peninsula, in the extreme northwest, make their living by selling handcrafted textiles and hammocks.

Colombia's cities have grown rapidly since the mid-1900's. The majority of the wealthy and middle-class people live in cities. The size of the middle class and the working class has grown as developing industries have

Most rural families in the Colombian Andes earn their living by growing crops on a small plot of land. They often live together in extended family groups, with several generations under one roof.

provided many new jobs. Middle-class Colombians live in comfortable houses or apartments. Many working-class people live in rundown buildings in older neighborhoods.

The cities have severe problems of poverty, unemployment, poor housing, and crime. Many unskilled rural laborers, seeking an escape from their backbreaking, low-paying work on the land, have poured into the nation's urban areas. There, they settle in huge, sprawling slums called *tugurios,* only to face unemployment or a low-paying job.

Most tugurios have no running water, electric power, or sewers. The impoverished settlers build shacks made of tin, cardboard, and other scrap materials. Some children in these areas run away or are abandoned by their parents, who cannot support them. These homeless children, called *gamines*, roam the streets and alleys.

The heart of Bogatá is the main square, called Plaza Bolívar. Santa Clara Cathedral and other historic buildings surround the plaza.

## SIMON BOLÍVAR

| | |
|---|---|
| 1783 | Born in Venezuela. |
| 1810 | Participates in throwing off Spanish rule in Caracas. |
| 1811 | Works for formal declaration of Venezuelan independence, which is made on July 5. |
| 1813 | Becomes dictator of Venezuela. |
| 1814 | Seizes Bogotá, Colombia. |
| 1819 | Defeats the Spaniards at Boyacá, Colombia, thereby liberating Colombia. Becomes president of Gran Colombia. |
| 1821 | Defeats Spaniards at Carabobo, Venezuela. |
| 1824 | Becomes dictator of Peru. |
| 1828 | Survives assassination attempt. |
| 1830 | Dies while leaving for Europe. |

Known as El Libertador (The Liberator), Simón Bolívar was one of South America's most successful generals. During the early 1800's, he and his army of patriots won independence for the modern nations of Bolivia, Colombia, Ecuador, Peru, and Venezuela. In 1819, he led a congress that formed the republic of Gran Colombia, uniting what are now Colombia and Venezuela. Panama joined in 1821 and Ecuador in 1822.

Bolívar became dictator of Peru in 1824. His army won a victory over the Spaniards at Ayacucho in 1824, which ended Spanish power in South America. Upper Peru became a separate state, named Bolivia in Bolívar's honor, in 1825.

Bolívar hoped to form a union of the new South American nations against Spain, having close relations with the United States. But his hopes were not achieved. By 1830, the republic of Gran Colombia had split into three separate countries.

Feelings against Bolívar grew violent. He resigned as president of Colombia in 1830.

# COLOMBIAN GOLD

In the 1530's, when Gonzalo Jiménez de Quesada led his expeditions into the highland valleys of the Andes, he and his fellow conquistadors had one goal: to find the legendary riches of El Dorado. Although the kingdom of El Dorado was never found, the quest for this mythic land of gold altered the course of South American history. Fired by greed and hungry for glory, the Spanish conquistadors marched boldly over the mountains and across the plains—and changed the face of Colombia forever.

The legend of El Dorado may have grown out of a coronation ceremony of the native Chibcha culture. During the ceremony, according to tradition, the new ruler was covered in gold dust and rowed out to the middle of the lake. It was said that the boat contained many objects made of gold and emeralds. According to legend, as the gold dust was washed off the new ruler, the priceless objects were cast into the lake as offerings to the gods.

Other indigenous people told stories about this custom to the Spanish conquistadors. The Spanish assumed that any land with such a custom must be full

Lake Guatavita is believed by some historians to be the site of the ceremony that inspired the legend of El Dorado. Beginning in 1562, several attempts were made to drain the lake, and gold objects were found buried in its muddy bottom. In 1965, the Colombian government prohibited all further privately sponsored "treasure hunts."

of treasures. They called the native ruler in the stories *El Dorado* (the Golden One). Later, the legendary kingdom also came to be called El Dorado.

Numerous expeditions into South American territory in search of El Dorado's wealth were led by explorers, including Gonzalo Pizarro, who left Quito (in present-day Ecuador) to strike out across the Andes, and Francisco de Orellana, who followed the course of the Napo and Amazon rivers.

By then, the Chibcha, whose coronation ceremony had created such a powerful myth, had been conquered by Jiménez de Quesada's armies. Their people were enslaved, their culture was devastated, and the Chibcha civilization was soon destroyed by the invaders' invincible greed.

All that remains of the Chibcha culture today is the magnificent goldwork that somehow escaped the Spanish melting pots. Now displayed in the Gold Museum in Bogotá, the beauty of these masterpieces is especially haunting because it was the Chibcha's remarkable talent as goldsmiths that helped bring about their downfall.

When the Spaniards first swarmed over the rocky hills into Chibcha territory, they found fortified towns, organized into two loose confederations. The wooden houses of the Chibcha were covered inside and out with clay and cane to keep out the mountain cold. Some—perhaps those of the wealthier families—even had cotton curtains in the windows and ornamental gold plates above the doorways.

The Chibcha were a farming people who cultivated fields of beans, corn, cotton, potatoes, and tobacco. They were also accomplished weavers and potters. But

Only 8 inches (20 centimeters) long, this small golden replica of a raft was found at the edge of Lake Siecha, one of the Chibcha's sacred lakes. It appears to depict the coronation ceremony that grew into the legend of El Dorado.

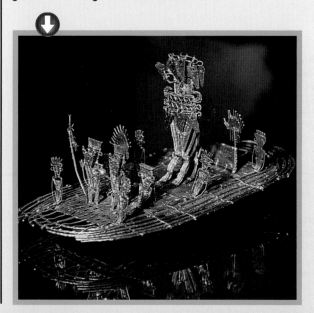

A present-day inhabitant of the Colombia highlands displays her gold teeth. The gold-working for which her ancestors were famous led early explorers to believe that the Andes contained large deposits of the precious metal.

Stunning artifacts from Bogotá's Gold Museum display the artistry of the Chibcha culture. The pieces show distinctive stylistic markings, including large ear disks.

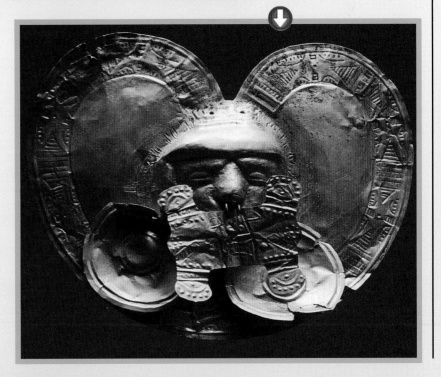

their greatest skill was their exquisite goldworking. They cast, welded, and hammered gold into magnificent earrings, necklaces, nose ornaments, masks, pendants, breastplates, bells, and bracelets.

The Chibcha also sculpted a variety of small figures out of pure gold, such as frogs, birds, snakes, and fish. Using the simplest tools, they created objects that even the artisans of today find difficult to reproduce.

The Chibcha way of life ended in the 1530's when, forced to defend their land against the Spanish invaders, they were quickly and brutally defeated. Tales of Chibcha chiefs weighed down by gold ornaments, face masks, and breastplates as they marched into battle—desperate to save their land and liberty—have kept the myth of El Dorado alive for hundreds of years.

# COMOROS

The small African nation of Comoros *(KOM uh ROHZ* or *kuh MAWR ohz)* consists of three main islands—Anjouan, Grande Comore, and Mohéli—and several smaller ones. A fourth main island—Mayotte—is claimed by Comoros, but the people of Mayotte have voted to remain a French possession.

The first people who lived on the Comoros Islands came from mainland Africa, Madagascar, and Malaysia. In the 1400's, Arabs took over the islands and ruled each one as a separate kingdom. France seized Mayotte in 1843 and had gained control of the rest of the island group by 1886

The French granted the Comoros Islands self-rule in 1961. In 1974, Anjouan, Grande Comore, and Mohéli voted for complete independence. But Mayotte voted to keep French rule then, and again in 1976.

After independence, a number of successful and unsuccessful coup attempts against elected governments occurred. In 1997, separatists on Anjouan and Mohéli each declared independence, claiming the central government had neglected their political and economic needs. The central government did not recognize their claims.

In April 1999, military leaders overthrew the elected government of Comoros. In December 2001, voters approved a new constitution that was designed to give greater autonomy to the three main islands of Comoros. The constitution provides for an elected national president, who comes from each of the three islands on a rotating basis, and a national assembly. Each island also has a president and a parliament. All three islands approved the constitution in 2002.

In 2007, Mohamed Bacar refused to step down as president of Anjouan at the end of his term. In 2008, the African Union, an organization of African nations that works for greater cooperation in Africa, and the government of Comoros sent troops to regain control of Anjouan.

Comoros is one of the world's poorest nations. It has no major industry, and no valuable minerals have been found there. The economy depends almost en-

## FACTS

| | |
|---|---|
| **Official name:** | Union des Comores (Union of the Comoros) |
| **Capital:** | Moroni |
| **Terrain:** | Volcanic islands, interiors vary from steep mountains to low hills |
| **Area:** | 719 mi² (1,862 km²) |
| **Climate:** | Tropical marine; rainy season (November to May) |
| **Main rivers:** | N/A |
| **Highest elevation:** | Mont Kartala, 7,746 ft (2,361 m) |
| **Lowest elevation:** | Indian Ocean, sea level |
| **Form of government:** | Independent republic |
| **Head of state:** | President |
| **Head of government:** | President |
| **Administrative areas:** | 3 islands with 4 municipalities |
| **Legislature:** | Assembly of the Union with 33 members serving five-year terms |
| **Court system:** | Cour Supremes (Supreme Court) |
| **Armed forces:** | N/A |
| **National holiday:** | Independence Day - July 6 (1975) |
| **Estimated 2010 population:** | 773,000 |
| **Population density:** | 1,075 persons per mi² (415 per km²) |
| **Population distribution:** | 72% rural, 28% urban |
| **Life expectancy in years:** | Male, 62; female, 66 |
| **Doctors per 1,000 people:** | 0.2 |
| **Birth rate per 1,000:** | 35 |
| **Death rate per 1,000:** | 8 |
| **Infant mortality:** | 68 deaths per 1,000 live births |
| **Age structure:** | 0-14: 42%; 15-64: 55%; 65 and over: 3% |
| **Internet users per 100 people:** | 3 |
| **Internet code:** | .km |
| **Languages spoken:** | Comorian, Arabic, French (all official) |
| **Religions:** | Sunni Muslim 98%, Roman Catholic 2% |
| **Currency:** | Comoran franc |
| **Gross domestic product (GDP) in 2008:** | $540 million U.S. |
| **Real annual growth rate (2008):** | 0.5% |
| **GDP per capita (2008):** | $758 U.S. |
| **Goods exported:** | Cloves, copra, vanilla, ylang-ylang (perfume oils) |
| **Goods imported:** | Cement, meat, petroleum products, rice, transportation equipment |
| **Trading partners:** | France, Pakistan, South Africa, Turkey, United Arab Emirates |

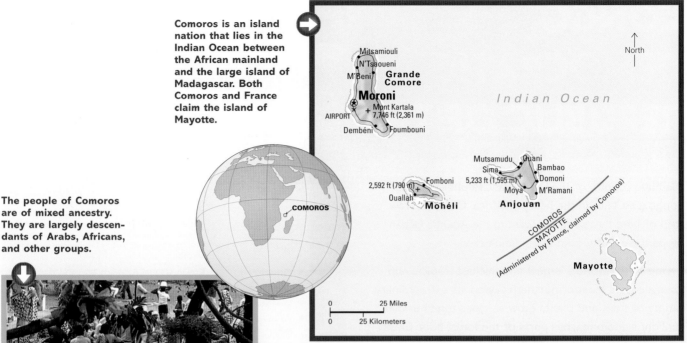

Comoros is an island nation that lies in the Indian Ocean between the African mainland and the large island of Madagascar. Both Comoros and France claim the island of Mayotte.

The people of Comoros are of mixed ancestry. They are largely descendants of Arabs, Africans, and other groups.

The beautiful islands of Comoros have many exotic plants and animals that are found nowhere else but nearby Madagascar. Most of the islands were formed by volcanic activity.

tirely on agriculture. About 85 percent of its workers farm for a living, raising such food crops as bananas, cassava, coconuts, corn, rice, and sweet potatoes. But good farmland is scarce, and the country must import large amounts of rice and other food. Comoros spends far more on imports than it earns from exports.

Most of the Comoros Islands were formed by volcanoes. Plateaus and valleys lie below the volcanic peaks. Mangrove swamps lie along almost all the shorelines. Heavy rains from November through April provide the islands with their only natural source of drinking water. The people store the rain water for use from May through October.

Hunger and disease are major problems on the Comoros Islands. Illness and malnutrition occur frequently due to poor diet, and the people suffer from a shortage of doctors and hospitals.

Most of the people of Comoros have mixed ancestry. They are descendants of Arabs, Africans, and other groups. Most Comorans are Muslims. Most people speak the Comorian language, which is an official language, along with Arabic and French.

The population lives mainly in rural villages. Moroni, the capital and largest city, lies on the west coast of Grande Comore Island. The city has about 50,000 people. Muslims from many countries come to visit Moroni's beautiful mosque.

# CONGO, DEMOCRATIC REPUBLIC OF THE

The huge country of the Democratic Republic of the Congo (DRC) lies in the heart of Africa. Only a narrow strip of land that stretches west to the Atlantic Ocean keeps the country from being landlocked.

One of the world's largest and thickest tropical rain forests covers most of northern Congo. Its extraordinary variety of trees and plants grow so close together that sunlight seldom reaches parts of the forest floor. Grassy savannas cover much of southern Congo. The country is sometimes called Congo (Kinshasa) to distinguish it from the neighboring Republic of the Congo, or Congo (Brazzaville). It was known as Zaire from 1971 to 1997.

## History

Ancestors of the Mbuti were the first known inhabitants of what is now Congo. They lived there in prehistoric times. Other Africans began moving into the area at least 2,000 years ago.

Beginning in the early 1500's, thousands of Africans in the area were enslaved and sold to the Portuguese and other Europeans. The slave trade ended in the early 1800's. In 1878, King Leopold II of Belgium hired a British explorer, Henry M. Stanley, to set up Belgian outposts along the Congo River. Leopold eventually gained control of the entire region and made it his personal colony, called the Congo Free State.

The people of the Congo Free State suffered horribly under Leopold's rule. Other countries protested. In 1908, the Belgian government responded by taking control from the king. Belgium refused to give the people any voice in their government. In 1959, rioting broke out against Belgian rule, and on June 30, 1960, the colony was granted independence. The new nation was named Congo. Civil war followed independence, as rival groups fought for power. In 1965, the Congolese army took control, and General Joseph Désiré Mobutu became president. In 1971, he changed the country's name to Zaire.

## FACTS

| | |
|---|---|
| • Official name: | Republique Democratique du Congo (Democratic Republic of the Congo) |
| • Capital: | Kinshasa |
| • Terrain: | Vast central basin is a low-lying plateau; mountains in east |
| • Area: | 905,535 mi² (2,344,858 km²) |
| • Climate: | Tropical; hot and humid in equatorial river basin; cooler and drier in southern highlands; cooler and wetter in eastern highlands; north of equator, the wet season is April to October, dry season December to February; south of equator, the wet season is November to March, dry season April to October |
| • Main rivers: | Congo, Ubangi, Aruwimi, Lomami, Kasai |
| • Highest elevation: | Margherita Peak, 16,762 ft (5,109 m) |
| • Lowest elevation: | Atlantic Ocean, sea level |
| • Form of government: | Dictatorship |
| • Head of state: | President |
| • Head of government: | Prime minister |
| • Administrative areas: | 25 provinces, 1 city |
| • Legislature: | Parliament consisting of the National Assembly with 500 members serving five-year terms and the Senate with 108 members serving five-year terms |
| • Court system: | Constitutional Court, Court of Cassation, Council of State |
| • Armed forces: | 139,300 - 151,300 troops |
| • National holiday: | Independence Day - June 30 (1960) |
| • Estimated 2010 population: | 69,963,000 |
| • Population density: | 77 persons per mi² (30 per km²) |
| • Population distribution: | 67% rural, 33% urban |
| • Life expectancy in years: | Male, 51; female, 55 |
| • Doctors per 1,000 people: | 0.1 |
| • Birth rate per 1,000: | 44 |
| • Death rate per 1,000: | 13 |
| • Infant mortality: | 92 deaths per 1,000 live births |
| • Age structure: | 0-14: 47%; 15-64: 50%; 65 and over: 3% |
| • Internet users per 100 people: | 0.4 |
| • Internet code: | .cd |
| • Languages spoken: | French (official), Lingala (a lingua franca trade language), Swahili, Kikongo, Tshiluba |
| • Religions: | Roman Catholic 50%, Protestant 20%, Kimbanguist 10%, Muslim 10%, other 10% |
| • Currency: | Congolese franc |
| • Gross domestic product (GDP) in 2008: | $11.59 billion U.S. |
| • Real annual growth rate (2008): | 8.0% |
| • GDP per capita (2008): | $179 U.S. |
| • Goods exported: | Cobalt, copper, crude oil, diamonds |
| • Goods imported: | Food, machinery, petroleum, transportation equipment |
| • Trading partners: | Belgium, France, South Africa, Zambia |

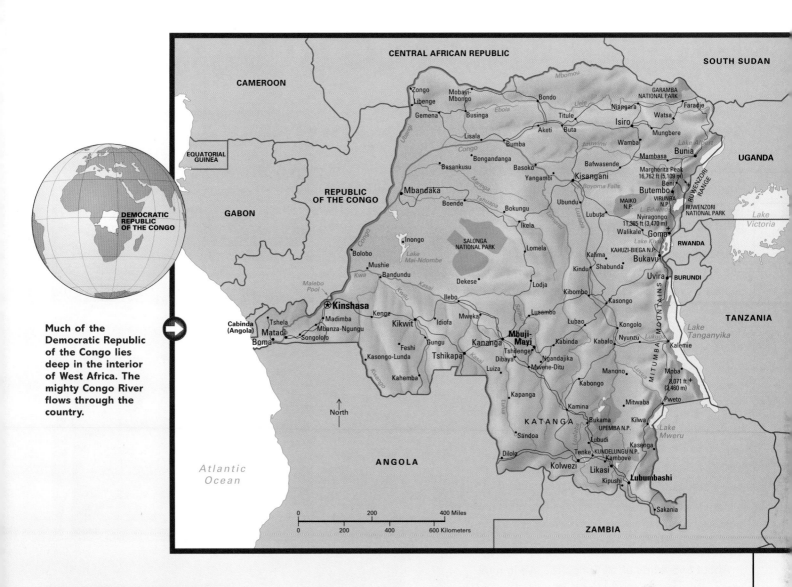

CENTRAL AFRICAN REPUBLIC

SOUTH SUDAN

CAMEROON

EQUATORIAL GUINEA

GABON

REPUBLIC OF THE CONGO

DEMOCRATIC REPUBLIC OF THE CONGO

UGANDA

RWANDA

BURUNDI

TANZANIA

ANGOLA

ZAMBIA

Atlantic Ocean

KATANGA

Much of the Democratic Republic of the Congo lies deep in the interior of West Africa. The mighty Congo River flows through the country.

North

0    200    400 Miles
0   200   400   600 Kilometers

## A troubled government

In 1991, public dissatisfaction with the country's lack of political freedom and growing economic problems forced Mobutu to allow opposition parties. Etienne Tshisekedi, an opponent of Mobutu, was elected prime minister in 1992, but Mobutu repeatedly tried to remove Tshisekedi from office.

In 1996, the government attempted to repatriate millions of Rwandans and many thousands of Burundians who were living in refugee camps. But fighting broke out as government troops fought refugee militias. Rebel militias opposed to the Mobutu regime also engaged government troops.

In May 1997, rebels led by Laurent Kabila marched into Kinshasa. Mobutu fled the country, and Kabila was sworn in as president on May 29. Kabila renamed the country the Democratic Republic of the Congo.

Violence continued. Congolese rebels, backed by Rwanda and Uganda, fought against Kabila's government, which received support from Angola, Namibia, and Zimbabwe. The United Nations sent peacekeeping forces in 2000, but fighting continued.

In 2001, Kabila was assassinated. Kabila's son, Joseph, succeeded him and tried to establish peace. Most foreign troops withdrew by the end of 2002, and Kabila arranged power-sharing agreements with his opponents. Still, some violence continued in eastern Congo.

In 2005, voters approved a new constitution, and in 2006, they voted in the first multiparty presidential and parliamentary election since 1965. Kabila won the election. He was reelected in 2011.

Since Mobutu's overthrow, Congo has suffered millions of conflict-related deaths, often from disease and malnutrition. Despite a number of cease-fires and power-sharing agreements, there has been no lasting peace.

# PEOPLE AND ECONOMY

When Congo became independent in 1960, Europeans greatly influenced the country's cultural life and economy. Deep divisions existed among the various ethnic groups, and the country faced severe economic problems. Since independence, Congo's leaders have worked to reduce European influence, unite the people, and improve the economy.

The people of Congo belong to many different ethnic groups. Tension between these groups has caused a great deal of conflict. The Mbuti, a Pygmy group, are descended from the first known inhabitants of the region. Most Congolese, however, are descended from African people who began to move into the area about 2,000 years ago.

Most of the country's ethnic groups have their own language that the people use in their everyday lives. About 200 languages are spoken, but most belong to the Bantu language group and thus are closely related. In addition, most Congolese speak one of the country's four regional languages—Kikongo, Lingala, Swahili, and Tshiluba. French is the nation's official language. It is used by government officials and taught in many schools.

About two-thirds of all Congolese live in rural areas, mainly in small villages that range from a few dozen to a few hundred people. Their houses are made from mud bricks or dried mud and sticks. Most of the homes have thatched roofs, but the houses of more well-to-do rural families have metal roofs. In some areas, people pound out rhythms on drums to send messages from village to village.

The great majority of rural families farm small plots of land and grow almost all their own food, including bananas, cassava, corn, peanuts, and rice. The basic Congolese dish is a porridge of grain and cassava, served with a spicy sauce.

A young miner at Kalima, in east-central Congo, removes tin from ore by repeated washings. Tin is just one of the country's many valuable mineral resources.

A thick blanket of mist covers the dense tropical rain forest of Congo. Because of the lack of open space, few people live in the forest, but its trees yield palm oil, rubber, and timber.

Crops raised for sale include cacao, coffee, cotton, and tea. Few farmers can afford modern equipment. As a result, production is low, and most farm families are poor.

Since independence, large numbers of Congolese—especially young people—have moved to urban areas seeking work. Today, about a third of the Congolese people live in cities. This rapid urban growth has caused such problems as unemployment and crowded living conditions. Government officials and business people live in attractive bungalows, but large numbers of urban factory and office workers are crowded into small, flimsy houses made of cinder blocks or mud bricks.

Congo is a poor country with a developing economy. Urban factory workers produce relatively small amounts of manufactured goods, mainly beer, cement, processed foods, soft drinks, steel, textiles, and tires. Many manufactured goods are imported.

Copper is the country's most important mineral resource. Congo ranks among the leading copper-producing nations and leads the world in producing industrial diamonds, its second most important mineral. It produces petroleum from deposits off the coast. The nation also has deposits of cadmium, cobalt, *coltan,* gold, manganese, silver, tin, and zinc. Coltan is a mineral used in electronic devices, such as cellular phones and laptop computers.

Civil war throughout the late 1990's brought Congo's economy almost to a standstill. Most exports stopped, the mining industry almost shut down, and most cities lacked electricity. Widespread malnutrition and a substantial increase in AIDS cases added to the woes of Congo's people. Sporadic outbreaks of violence continued through the 2000's.

**Rwandan troops ride through a village in the eastern Congo. Starting in the late 1990's, the country was the battleground in a brutal civil war involving rival ethnic militias from Congo, Rwanda, and Uganda.**

# THE CONGO RIVER

The fifth longest river in the world, the Congo River flows 2,900 miles (4,667 kilometers) through the heart of Africa. Carrying more water than any other river except the Amazon, it drains an area of about 1.4 million square miles (3.6 million square kilometers). The Congo River is the main waterway of the Democratic Republic of the Congo.

The first European to see the river was Portuguese navigator Diogo Cão, who reached its mouth in 1483. Portuguese settlers established an outpost on the Congo's southern bank near the Atlantic Ocean in the 1490's. But Europeans knew little about the rest of the great river until after the British explorer Henry M. Stanley completed an expedition from its source to its mouth in 1877.

The Congo begins south of Kabalo, Democratic Republic of the Congo, where the Lualaba and Luvua rivers meet. The river is often called the Lualaba from this point until it tumbles over Stanley Falls, where it becomes the Congo.

Near Stanley Falls, the river turns westward and flows through the rain forest of northern Democratic Republic of the Congo, where several major rivers empty into the Congo, including the Aruwimi, Lomami, and Ubangi. Near the town of Mbandaka, the Congo turns southwestward to form a natural boundary between the countries of the Democratic Republic of the Congo and the Republic of the Congo for about 500 miles (800 kilometers).

Near Kinshasa, the Congo widens so much that it forms a lake called Stanley Pool. The river then drops about 800 feet (240 meters), forming a series of spectacular waterfalls between Kinshasa and Matadi. These falls prevent riverboats from sailing all the way to the Atlantic Ocean.

The Congo empties into the Atlantic about 90 miles (140 kilometers) west of Matadi. Unlike the Mississippi and the Nile, the Congo does not form a delta at its mouth. Instead, the river's muddy waters flow into a deep trench that extends far into the ocean.

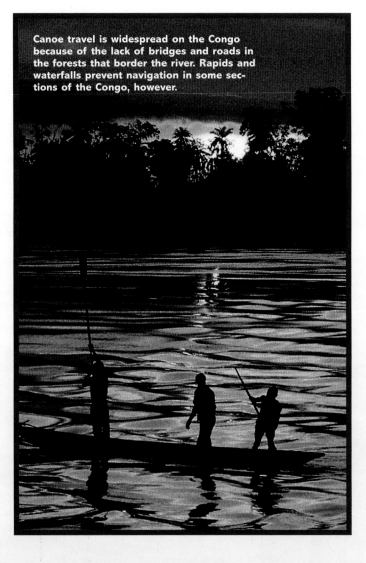

Canoe travel is widespread on the Congo because of the lack of bridges and roads in the forests that border the river. Rapids and waterfalls prevent navigation in some sections of the Congo, however.

Commercial ships sail the Congo between the Atlantic and Matadi and between Kinshasa and Kisangani. The river also serves as a major transportation route for local people. Fishing ranks as the most important economic activity in all areas of the Congo River Basin, but little agricultural activity is possible in the dense forest.

People have settled in areas where the riverbanks are relatively firm and permanent. But elsewhere, swampy conditions and the possibility of floods make living conditions much more difficult. In the densely forested, hard-to-reach northern areas of the Congo River Basin, Pygmies carry on their traditional ways, hunting and gathering food as they travel in small groups from one area to another.

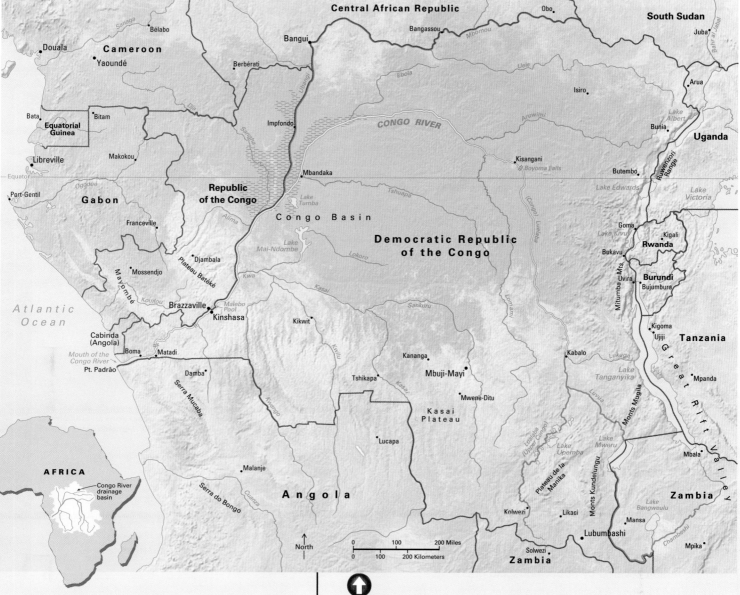

The map shows the Congo River drainage basin in central Africa, including the following countries and features:

**Central African Republic** — Obo, Bangassou, Bangui, Berbérati, Ebola

**South Sudan** — Juba, Arua, Bahr al Jabal

**Cameroon** — Douala, Yaoundé, Sanaga, Bélabo

**Equatorial Guinea** — Bata, Bitam

**Gabon** — Libreville, Port-Gentil, Franceville, Makokou, Mossendjo, Ogooué, Mayombe, Djoué

**Republic of the Congo** — Impfondo, Mbandaka, Brazzaville, Djambala, Plateau Batéké, Alima, Koulou, Sangha, Ubangi

**Democratic Republic of the Congo** — Kinshasa, Kisangani, Boyoma Falls, Butembo, Bunia, Goma, Bukavu, Kikwit, Kananga, Mbuji-Mayi, Mwene-Ditu, Tshikapa, Lucapa, Kabalo, Kolwezi, Likasi, Lubumbashi, Boma, Matadi, Damba, Congo Basin, Lake Mai-Ndombe, Lake Tumba, Kasai Plateau, Plateau de la Manika, Monts Kundelungu, Lake Upemba, Lake Mweru, Lualaba, Lomami, Sankuru, Lokoro, Tshuapa, Aruwimi, Uele, Lukenie, Kwango, Kwilu, Luvua, Lukuga

**Uganda** — Ruwenzori Range, Lake Albert, Lake Edwards

**Rwanda** — Kigali, Lake Kivu

**Burundi** — Bujumbura

**Tanzania** — Kigoma, Ujiji, Mpanda, Mbala, Great Rift Valley, Lake Tanganyika, Monts Mogila

**Zambia** — Solwezi, Mansa, Mpika, Lake Bangweulu, Chambeshi

**Angola** — Malanje, Serra Mucaba, Serra do Bongo, Cuanza, Cabinda (Angola), Mouth of the Congo River, Pt. Padrão

Atlantic Ocean, Equator, Mbomou, Lake Victoria, Mittumba Mts., Uvira, Kabale

North
0   100   200 Miles
0   100   200 Kilometers

AFRICA — Congo River drainage basin

The Congo River drains a vast area of equatorial Africa. Despite its many waterfalls and rapids, the river is Congo's main waterway and an important transportation route. Oceangoing ships use the river between the Atlantic and Matadi, and other commercial ships navigate its waters between Kinshasa and Kisangani. Goods are transported between Kinshasa and Matadi by railroad.

The gorilla, the world's largest ape and one of its most intelligent and peaceful creatures, survives in the dense rain forests of the Congo River Basin. An adult male gorilla may weigh as much as 450 pounds (204 kilograms).

The rain forests that lie along the mighty Congo are also home to a remarkable variety of wild animals. Crocodiles and hippopotamuses live in or near the river, while baboons, chimpanzees, gorillas, and monkeys thrive in the forests. The okapi, a forest-dwelling animal related to the giraffe, lives nowhere else in the world but the Congo River Basin. The okapi was unknown to Europeans until the year 1900, more than 400 years after the first European saw the river itself.

# CONGO, REPUBLIC OF THE

The hot, humid Republic of the Congo lies in west-central Africa, directly on the equator. It takes its name from the Congo River, which flows along much of its eastern border. Its name in French, the official language, is République du Congo. The country is sometimes called Congo (Brazzaville) to distinguish it from neighboring Democratic Republic of the Congo, or Congo (Kinshasa).

## Land and economy

Thick tropical rain forests of tall trees, tangled bushes, and lush vines cover the northern half of Congo. Many exotic animals live in this section of the country, which forms part of the Congo River Basin. South of the river basin, in central Congo, is the Batéké Plateau, a grassy plain divided by deep, forested valleys. Still farther south lies Stanley Pool, a lake formed where the Congo River widens. Brazzaville, Congo's capital, is located on Stanley Pool.

To the west of Stanley Pool is the Niari Valley, a farming region covered by both woods and grassland. At the Mayombé Escarpment, a series of ridges and plateaus west of the valley, the land drops down to the Coastal Plain.

Building and maintaining roads in Congo is difficult because of the heavy rains and thick forests. Yet the country has one of the longest transportation systems in Africa, mainly because of the Congo-Ubangi river system. Boats carry goods between countries to the north and Stanley Pool. At that point, rapids prevent boats from getting to the ocean, so the Congo-Ocean railroad was built to link Brazzaville on Stanley Pool with Pointe-Noire on the Atlantic.

Congo's economy is based mainly on petroleum production, forestry, and agriculture. Petroleum is the country's most valuable resource and chief export. Lumber from the rain forests is the second

## FACTS

| | |
|---|---|
| Official name: | Republique du Congo (Republic of the Congo) |
| Capital: | Brazzaville |
| Terrain: | Coastal plain, southern basin, central plateau, northern basin |
| Area: | 132,047 mi² (342,000 km²) |
| Climate: | Tropical; rainy season (March to June); dry season (June to October); constantly high temperatures and humidity; particularly hot and humid along the equator |
| Main rivers: | Congo, Ubangi, Kouilou |
| Highest elevation: | 3,412 ft (1,040 m) near Gabon border |
| Lowest elevation: | Atlantic Ocean, sea level |
| Form of government: | Republic |
| Head of state: | President |
| Head of government: | President |
| Administrative areas: | 10 regions, 1 commune |
| Legislature: | Parliament consisting of the senate with 66 members serving five-year terms and the National Assembly with 137 members serving five-year terms |
| Court system: | Cour Supreme (Supreme Court) |
| Armed forces: | 10,000 troops |
| National holiday: | Independence Day - August 15 (1960) |
| Estimated 2010 population: | 4,012,000 |
| Population density: | 30 persons per mi² (12 per km²) |
| Population distribution: | 61% urban, 39% rural |
| Life expectancy in years: | Male, 52; female, 55 |
| Doctors per 1,000 people: | 0.2 |
| Birth rate per 1,000: | 37 |
| Death rate per 1,000: | 12 |
| Infant mortality: | 79 deaths per 1,000 live births |
| Age structure: | 0-14: 42%; 15-64: 55%; 65 and over: 3% |
| Internet users per 100 people: | 4 |
| Internet code: | .cg |
| Languages spoken: | French (official), Lingala and Monokutuba (lingua franca trade languages), many local languages and dialects with Kikongo having the most users |
| Religions: | Christian 50%, animist 48%, Muslim 2% |
| Currency: | Coopération Financière en Afrique Centrale franc |
| Gross domestic product (GDP) in 2008: | $10.77 billion U.S. |
| Real annual growth rate (2008): | 8.1% |
| GDP per capita (2008): | $2,748 U.S. |
| Goods exported: | Mostly: petroleum Also: lumber, sugar |
| Goods imported: | Food, iron and steel, machinery, pharmaceuticals, vehicles |
| Trading partners: | China, France, South Korea, United States |

most important export. Farmers raise cassava, yams, sweet potatoes, corn, and other crops to feed their families and to trade. Large plantations in the fertile Niari Valley produce palm oil and sugar cane.

## People

Congo's people belong to four main African ethnic groups. About 45 percent are Kongo farmers who live in the west and southwest. About 20 percent are Batéké hunters and fishermen who live north of Brazzaville, and about 10 percent are M'Bochi people, who live in the north and work mainly as clerks and technicians. The fourth major group, the Sangha, live in the northern forest.

## Politics

Congo came under French protection in 1880 and was made part of French Equatorial Africa in 1910. It became an independent nation on Aug. 15, 1960. In 1969, a military coup brought a group of army officers to power. They established a one-party Marxist state, which continued until 1991, when opposition political parties were legalized.

In 1992, voters approved a new constitution and elected Pascal Lissouba as president. But political and ethnic strife soon broke out. Tensions developed between Lissouba and two rivals—General Denis Sassou-Nguesso, who had led Congo as dictator from 1979 to 1992, and Bernard Kolelas.

In 1997, rebels led by Sassou-Nguesso overthrew Lissouba's government, and Sassou-Nguesso declared himself president. In 1998, delegates from many of the country's political parties elected 75 members to a National Transitional Council to act as the country's legislature until national elections could be held.

In 2002, voters approved a new constitution calling for an elected president and a two-house parliament. Later that year, Sassou-Nguesso was elected president under the new constitution. Most fighting ended in 2003. Sassou-Nguesso was reelected in 2009.

A gold-mining camp on the Mayombé Escarpment consists of tin-roofed shacks amid dense tropical woods. Congo has poor soil but several mineral resources, including copper, gold, lead, petroleum, potash, uranium, and zinc.

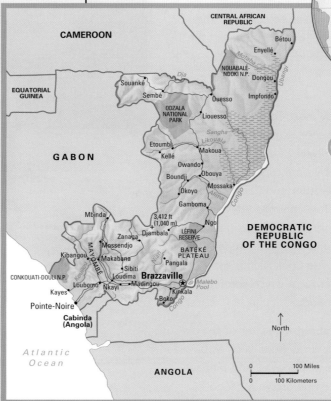

The Republic of the Congo is situated on the western and northern banks of the Congo River. The country is open to the Atlantic Ocean between Gabon, on the north, and the Democratic Republic of the Congo and Cabinda (Angola) on the south.

# COOK ISLANDS

The 15 islands of the Cook group are spread out over 850,000 square miles (2.2 million square kilometers) in the South Pacific Ocean, about 1,800 miles (2,900 kilometers) northeast of New Zealand. The Cook Islands form two groups—the Southern Group and the Northern Group. Most of the islands of the Southern Group are raised volcanic islands. Rarotonga, the principal island of the Cooks, is in the Southern Group. Most islands of the Northern Group are coral atolls.

## People

The Cook Islanders, who are mainly Polynesians, call themselves *Maori*. Their language is closely related to that of the New Zealand Maori, as are many of their customs. The Cook Islands have a population of about 21,000. The only large settlement, Avarua on Rarotonga, is also the capital and commercial center. The official languages are English and Cook Island Maori.

Attending church services is important to the people of the Cook Islands. The majority of islanders belong to the Cook Islands Christian Church. This church follows the teachings of the former London Missionary Society, which sent the first missionaries to the islands in the 1820's.

## Economy

Service industries employ most Cook Island workers. The most important service industry is tourism. Many visitors come from New Zealand, Australia, and Europe. Some islanders are concerned about the impact of so many visitors, and Cook Island government policy calls for "controlled development" of tourism to preserve the landscape and the environment. Waste management and fuel consumption are problems. The Cook Islands must import petroleum and other fuels, which accounts for about a quarter of all imports.

Agriculture employs about a quarter of the Cook Islanders. Farmers grow fruits such as bananas, coconuts, mangoes, papayas, and pineapples to sell. Additional crops grown for food

Wearing the traditional grass skirt, an islander makes music on a slit-drum. Many traditional customs have disappeared from the Cook Islands as a result of European influence, but some of the old ways persist.

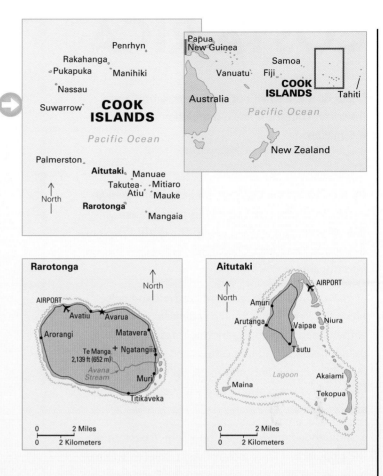

The Cook Islands sprawl over a wide expanse of the South Pacific. The largest of the 15 islands is Rarotonga in the Southern Group.

A jagged, rocky outcrop rises from lush, coastal vegetation in the Southern Group of the Cook Islands. Most of the Southern Group consists of raised volcanic islands whose fertile soil yields rich harvests.

Cook Islanders assemble for church. The majority of islanders are members of the Cook Islands Christian Church. This church follows the teachings of the London Missionary Society, which introduced Christianity to the islands in the 1820's.

include cassava and sweet potatoes. Pearl oyster farming is an important industry. The main exports of the Cook Islands are pearls and fish, especially tuna.

Manufacturing in the Cook Islands is limited. The main products are beer, clothing, handicrafts, and processed fruit. Mines produce cobalt, copper, manganese, and nickel.

## History and government

In 1773, Captain James Cook became the first known European to reach the islands. The United Kingdom took control of the islands in 1888 and gave administrative control to New Zealand in 1891. A new constitution gave the islanders control of their internal affairs in 1965.

Today, the islands have an arrangement with New Zealand called *free association*. Under free association, the Cooks are self-governing, but the people are citizens of New Zealand. That nation offers the islands military support for defense and provides economic assistance. In recent years, a large number of Cook Islanders have migrated to New Zealand.

A New Zealand representative, called the high commissioner, lives on Rarotonga and handles that nation's relations with the Cook Islands. The head of state is the British monarch, who also has a representative living on Rarotonga. The islands have a 24-member Parliament, which chooses the prime minister. The prime minister appoints a cabinet.

# COSTA RICA

Costa Rica is a small but rapidly growing Central American country. It has a population of about 4.6 million.

## Ancestry

By A.D. 1000, the Corobici, an *indigenous* (native) American people, had settled in the northern valleys of what is now Costa Rica. The Boruca lived in the south. The Carib, Chorotega, and Nahua moved into the area in the 1400's.

In the 1500's, the Spaniards arrived and colonized the land. Many Spanish colonists intermarried with indigenous people, and their descendants are called *mestizos*. Today, mestizos and people of unmixed European ancestry make up the majority of Costa Rica's population. Nearly all Costa Ricans speak Spanish.

Indigenous groups make up about 1 percent of the population. They mostly live in isolated communities in the highlands and along the Pacific and Caribbean shores. They follow their ancestors' traditional ways of life.

About 3 percent of the population is of African descent. Most of these residents live along the Caribbean coast. Their ancestors originally came to Costa Rica from the Caribbean island of Jamaica in the late 1800's to build railroads and work on banana plantations. Many of them speak a Jamaican *dialect* (local form) of English.

## The economy

About a third of Costa Rica's people live on farms or in rural towns, and about a seventh of its workers are employed in farming or ranching. Costa Rica's most valuable natural resource is its fertile volcanic soil.

Poás volcano is easily accessible to tourists with a paved road leading up to its smoky crater. Such volcanoes help make the soil of Costa Rica rich and fertile.

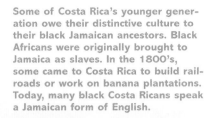

Coffee plants produce Costa Rica's leading export. Hillsides covered with coffee plants surround San José, the capital. Coffee beans, bagged in 132-pound (60-kilogram) burlap sacks are ready for shipping from the *beneficio* (processing plant) to the roasting plant. There, they are blended, roasted, ground, and packed for market.

The chief agricultural products are bananas, beef cattle, *cacao* (seeds used to make chocolate), coffee, corn, ornamental plants, pineapple, poultry products, rice, and sugar cane. Farmers grow a wide variety of fruits and vegetables, including oranges, beans, and potatoes.

Manufacturing employs about a seventh of the nation's workers, and it is growing in importance. The leading manufactured products include clothing, computer chips, cosmetics, fertilizer, medical equipment, medicines, processed foods, and textiles.

Rich forests of oaks, pines, and tropical hardwoods such as mahogany and cedrela cover much of the land. Costa Rica also has small deposits of bauxite and manganese. In addition, many foreign tourists, who come to enjoy the sunshine and beauty of the land, also help boost the economy.

Costa Rica's economy depends heavily on foreign exports. Traditionally, agricultural products dominated the export trade. Today, the main agricultural exports are bananas, coffee, and pineapples. Since the late 1990's, computer chips and medical equipment manufactured in San José have also become major exports.

In 2007, Costa Rica ratified the Dominican Republic-Central America-United States Free Trade Agreement (CAFTA-DR), designed to reduce trade barriers among the agreement's members. The treaty went into effect in Costa Rica on Jan. 1, 2009.

# COSTA RICA TODAY

Costa Rica is a land of rugged mountains, fertile soil, and lush forests. Its name, meaning *rich coast* in Spanish, came from explorers who heard tales of precious metals in the region. However, the land has little mineral wealth.

## The land

High mountain ranges called *cordilleras* cross Costa Rica from northwest to southeast and divide the country into three land regions.

In the Central Highlands lie two large, fertile areas—the *Meseta Central* (Central Plateau) and the *Valle del General* (Valley of the General). Steep cordilleras surround each area. The Meseta Central is the heartland of Costa Rica. About 75 percent of Costa Rica's people live there, and the capital, San José, stands on the plateau. The Meseta's rich volcanic soil and favorable climate make it the country's chief coffee-growing region. The Valle del General lies to the southeast of the Meseta. Slightly warmer and wetter than the Meseta, it is an agricultural region of hills and plains.

The Caribbean Lowlands form a second major land region of Costa Rica. This wide band of swampy tropical jungles lies along the east coast.

The Pacific Coastal Strip is a region of lowlands along the west coast. The strip has an ideal climate for growing bananas.

## The government

Costa Rica is a democratic republic. The people elect a president to a four-year term. The president, the two vice-presidents, and the members of the Cabinet make up the Council of Government. The council, which conducts foreign affairs and enforces federal laws, may also veto bills passed by

# FACTS

| | |
|---|---|
| Official name: | Republica de Costa Rica (Republic of Costa Rica) |
| Capital: | San José |
| Terrain: | Coastal plains separated by rugged mountains |
| Area: | 19,730 mi² (51,100 km²) |
| Climate: | Tropical and subtropical; dry season (December to April); rainy season (May to November); cooler in highlands |
| Main rivers: | General, San Carlos, Chirripó |
| Highest elevation: | Chirripó Grande, 12,530 ft (3,819 m) |
| Lowest elevation: | Pacific Ocean, sea level |
| Form of government: | Democratic republic |
| Head of state: | President |
| Head of government: | President |
| Administrative areas: | 7 provincias (provinces) |
| Legislature: | Asamblea Legislativa (Legislative Assembly) with 57 members serving four-year terms |
| Court system: | Corte Suprema (Supreme Court) |
| Armed forces: | None |
| National holiday: | Independence Day - September 15 (1821) |
| Estimated 2010 population: | 4,672,000 |
| Population density: | 237 persons per mi² (91 per km²) |
| Population distribution: | 62% urban, 38% rural |
| Life expectancy in years: | Male, 75; female, 81 |
| Doctors per 1,000 people: | 1.3 |
| Birth rate per 1,000: | 17 |
| Death rate per 1,000: | 4 |
| Infant mortality: | 10 deaths per 1,000 live births |
| Age structure: | 0-14: 27%; 15-64: 67%; 65 and over: 6% |
| Internet users per 100 people: | 34 |
| Internet code: | .cr |
| Languages spoken: | Spanish (official), English |
| Religions: | Roman Catholic 70.5%, Evangelical Protestant 13.8%, other 15.7% |
| Currency: | Costa Rican colón |
| Gross domestic product (GDP) in 2008: | $30.01 billion U.S. |
| Real annual growth rate (2008): | 3.0% |
| GDP per capita (2008): | $6,596 U.S. |
| Goods exported: | Bananas, coffee, computer chips, medical equipment, melons, ornamental plants, pineapples, seafood, sugar |
| Goods imported: | Chemicals, machinery, petroleum, transportation equipment |
| Trading partners: | China, Mexico, Netherlands, United States |

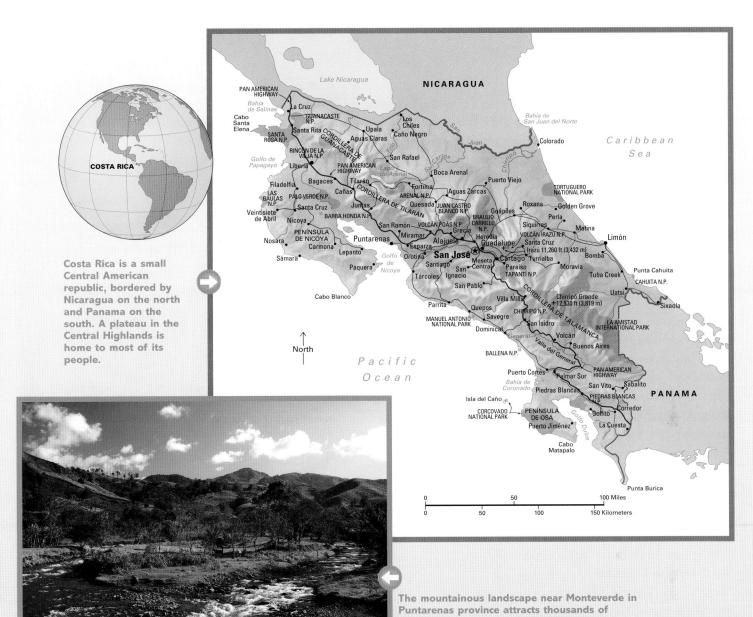

Costa Rica is a small Central American republic, bordered by Nicaragua on the north and Panama on the south. A plateau in the Central Highlands is home to most of its people.

The mountainous landscape near Monteverde in Puntarenas province attracts thousands of tourists every year who visit the area to enjoy the rich variety of plant and animal life.

the 57-member Legislative Assembly. The people elect the legislators to four-year terms.

The Costa Rican Constitution was adopted in 1949. Like other Central American countries, Costa Rica has a history of Spanish colonization, followed by a period of dictatorships and revolts. The region was a Spanish colony for about 300 years after Christopher Columbus landed in Costa Rica in 1502. In 1821, Costa Rica and other Central American states broke away from Spain. They became part of Mexico, but in 1823 formed the United Provinces of Central America. When the union began to collapse in 1838, Costa Rica declared its independence and fell under the dictatorship of Braulio Carrillo.

But Costa Rica has also had strong, effective leaders. From 1849 to 1859, President Juan Rafael Mora established the first national bank, first streetlight system, and many public schools. Under Julio Acosta, who became president in 1919, and his successors, Costa Rica became a model of democracy and social reform.

Since 1974, Costa Rica has had an orderly succession of democratic governments. President Oscar Arias Sánchez, who served from 1986 to 1990, won the 1987 Nobel Peace Prize for his Central American peace plan. Arias was reelected in 2006.

In 2010, Laura Chinchilla became the first woman president of Costa Rica.

# TOURISM

The gold-hungry Spanish explorers who gave Costa Rica the name *rich coast* found neither gold nor silver. But today, Costa Rica offers riches of another kind. Costa Rica's favorable climate, natural beauty, and culture attract hundreds of thousands of tourists every year.

## Cultural attractions

San José is Costa Rica's bustling capital and largest city. Lying in a valley in the mountainous interior, the city has a mild climate, with an average temperature of about 73° F (23° C).

San José is a picturesque mix of old and new—with Spanish-style churches and houses standing among modern stores and office buildings. The National Theater and Gran Hotel Costa Rica dominate the Culture Plaza—a large, beautiful square in the capital. The city's National Monument is an impressive statue dedicated to the five Central American republics who ousted William Walker, an American who tried to take control of Nicaragua in the 1800's.

Colorful festivals on religious holidays are also a tourist attraction. During the annual Christmas festivals in San José, for example, thousands of tourists as well as Costa Ricans come to enjoy the bullfights, fireworks, and parades of masked merrymakers.

## Natural attractions

Many tourists also come to enjoy the natural beauty of Costa Rica. Rugged mountains and lush tropical forests cover most of the land, and warm ocean waters roll onto its sandy beaches.

Many national parks have been created to preserve these natural wonders. The country's park system includes beaches where sea turtles come to lay their eggs, tropical rain forests that provide a habitat for chattering monkeys and colorful birds, and several active volcanoes.

Cocos Island (Isla del Coco) is a small island that lies about 200 miles (320 kilometers) southwest of Costa Rica

Cocos Island, off the Costa Rican mainland in the Pacific Ocean, is home to unusual species of birds. To preserve their habitat, Costa Rica has made the island a national park.

Costa Rica has an extensive national park system, which provides enjoyment for tourists and residents alike.

Guanacaste
Santa Rosa
Rincón de la Vieja
Volcán Tenorio
Las Baulas
Palo Verde
Arenal
Tortuguero
Barra Honda
Juan Castro Blanco
Braulio Carrillo
Volcán Poás
Volcán Turrialba
Volcán Irazú
Barbilla
Cahuita
Carara
Tapantí
Chirripó
Manuel Antonio
La Amistad International Park
Ballena
Piedras Blancas
Corcovado

Colorful religious festivals reflect Costa Rica's Roman Catholic heritage. The festivals are also major tourist attractions.

in the Pacific Ocean. The entire island is a national park. Rare sea birds nest there, and waterfalls spill out of the evergreen forest into the sea.

Cahuita National Park, south of Limón on the Caribbean coast, is one of the parks that preserve the rain forest. Like rain forests throughout the world, the rain forests of Costa Rica are a precious natural resource with a wealth of plant and animal life. A tropical rain forest has more kinds of trees than any other area in the world. In addition, more species of amphibians, birds, insects, mammals, and reptiles live in tropical forests than anywhere else.

In the Central American rain forest of Costa Rica, squirrel monkeys scamper along tree branches and climb vines. Glittering emerald-green-and-crimson birds called *quetzals* perch high in the trees. And *jaguars*—large, powerful wild cats—live and hunt in the forests too.

Poás and Irazú are two active volcanoes in the Cordillera Central of Costa Rica—the mountain range that rings the plateau where San José lies. Tourists and scientists alike can view the two volcanoes at close range.

A rich variety of wildlife makes its home in Costa Rica's national parks, providing enjoyment for tourists and residents alike.

1 Quetzal
2 Spider monkey
3 Pygmy anteater
4 Squirrel monkey
5 Mouse opossum
6 Jaguar
7 Agrias scardanapalus
8 Nessaea obrinus
9 Atlantic green turtle
10 Tarantula
11 Alligator

République de Côte d'Ivoire *(koht dee VWAR)* lies along the Gulf of Guinea on the western bulge of Africa. The land rises gradually from the Atlantic Ocean to a tropical forest in the interior. In the north, the forest changes to grassland with scattered trees.

Low, rocky cliffs line the southwestern coast of the country, while the southeastern coast is flat and sandy. A sand bar that runs for 180 miles (289 kilometers) along this section of the coast is bordered by deep lagoons.

Yamoussoukro is the capital of Côte d'Ivoire. However, most government offices are in Abidjan, the administrative center and former capital of the country. Abidjan is also Côte d'Ivoire's largest city and main port.

Côte d'Ivoire, also known as Ivory Coast, received its name from French sailors who came to the region in the late 1400's to trade for ivory. Before that time, great African kingdoms had ruled there. After the Europeans arrived, the slave trade became important.

In 1842, the French took control of the area on the coast around Grand-Bassam, and later they made treaties with African chiefs to expand the area under French protection. Côte d'Ivoire became a French colony in 1893.

After World War I (1914-1918), France built ports, roads, and railroads in Côte d'Ivoire. After World War II (1939-1945), the French began developing the region's resources, and Côte d'Ivoire became the richest French colonial region in western Africa. France made Côte d'Ivoire a territory in the French Union in 1946.

In 1958, Côte d'Ivoire voted to become a self-governing republic within the French Community, an organization that linked France with its overseas territories. On August 7, 1960, Côte d'Ivoire declared itself an independent republic, but the nation kept close economic ties with France.

## FACTS

| | |
|---|---|
| Official name: | République de Côte d'Ivoire (Republic of Côte d'Ivoire) |
| Capital: | Yamoussoukro |
| Terrain: | Mostly flat to undulating plains; mountains in northwest |
| Area: | 124,504 mi$^2$ (322,463 km$^2$) |
| Climate: | Tropical along coast, semiarid in far north; three seasons—warm and dry (November to March), hot and dry (March to May), hot and wet (June to October) |
| Main rivers: | Bandama, Cavally, Komoé, Sassandra |
| Highest elevation: | Mount Nimba, 5,748 ft (1,752 m) |
| Lowest elevation: | Gulf of Guinea, sea level |
| Form of government: | Republic |
| Head of state: | President |
| Head of government: | Prime minister |
| Administrative areas: | 19 regions |
| Legislature: | Assemblee Nationale (National Assembly) with 225 members serving five-year terms |
| Court system: | Cour Supreme (Supreme Court) |
| Armed forces: | 17,100 troops |
| National holiday: | Independence Day - August 7 (1960) |
| Estimated 2010 population: | 21,059,000 |
| Population density: | 169 persons per mi$^2$ (65 per km$^2$) |
| Population distribution: | 52% rural, 48% urban |
| Life expectancy in years: | Male, 52; female, 55 |
| Doctors per 1,000 people: | 0.1 |
| Birth rate per 1,000: | 35 |
| Death rate per 1,000: | 14 |
| Infant mortality: | 89 deaths per 1,000 live births |
| Age structure: | 0-14: 41%; 15-64: 56%; 65 and over: 3% |
| Internet users per 100 people: | 2 |
| Internet code: | .ci |
| Languages spoken: | French (official), 60 native dialects with Dioula the most widely spoken |
| Religions: | Muslim 38.6%, Christian 32.8%, indigenous beliefs 11.9%, other 16.7% |
| Currency: | Communaute Financiere Africaine franc |
| Gross domestic product (GDP) in 2008: | $23.57 billion U.S. |
| Real annual growth rate (2008): | 2.7% |
| GDP per capita (2008): | $1,173 U.S. |
| Goods exported: | Cacao beans, coffee, cotton, crude oil and petroleum products, fish, fruits |
| Goods imported: | Fish, fuel, rice and other food, vehicles |
| Trading partners: | France, Germany, Italy, Netherlands, United States |

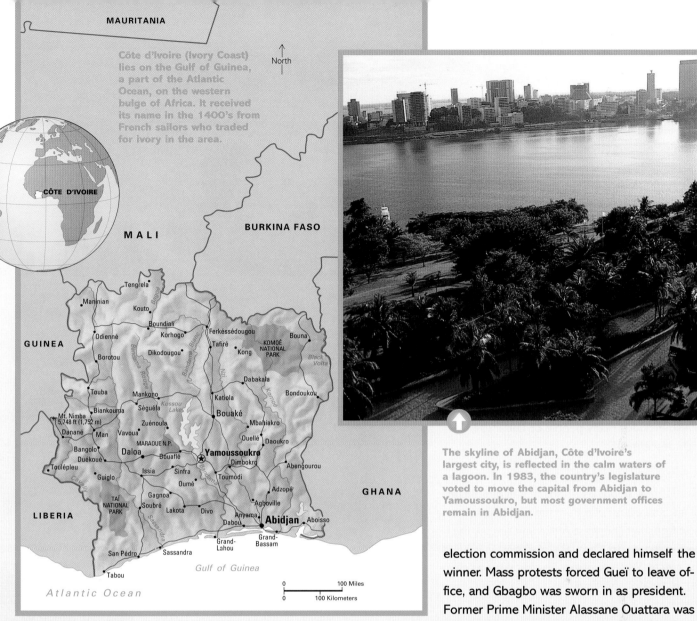

Côte d'Ivoire (Ivory Coast) lies on the Gulf of Guinea, a part of the Atlantic Ocean, on the western bulge of Africa. It received its name in the 1400's from French sailors who traded for ivory in the area.

North

MAURITANIA

CÔTE D'IVOIRE

MALI

BURKINA FASO

GUINEA

Tengrela

Maninian

Kouto

Boundiali

Odienné

Korhogo

Ferkèssédougou

Bouna

Borotou

Dikodougou

Tafiré

Kong

KOMOÉ NATIONAL PARK

Black Volta

Dabakala

Touba

Mankono

Katiola

Bondoukou

Biankouma

Séguéla

Kossou Lake

Bouaké

Mt. Nimba 5,748 ft (1,752 m)

Zuénoula

Mbahiakro

Danané

Vavoua

Ouellé

Daoukro

Man

MARAOUE N.P.

Bangolo

Daloa

Bouaflé

Yamoussoukro

Dimbokro

Abengourou

Duékoué

Issia

Sinfra

Toumodi

Touléplleu

Guiglo

Oumé

Adzopé

TAÏ NATIONAL PARK

Gagnoa

Soubré

Lakota

Divo

Agboville

GHANA

LIBERIA

Anyama

Abidjan

Aboisso

Dabou

San Pédro

Sassandra

Grand-Lahou

Grand-Bassam

Tabou

Gulf of Guinea

Atlantic Ocean

0   100 Miles
0   100 Kilometers

The skyline of Abidjan, Côte d'Ivoire's largest city, is reflected in the calm waters of a lagoon. In 1983, the country's legislature voted to move the capital from Abidjan to Yamoussoukro, but most government offices remain in Abidjan.

Félix Houphouët-Boigny, who had been a leader of the movement for independence among French territories in western Africa, was elected president of Côte d'Ivoire for seven 5-year terms beginning in 1960. He helped unite the country's many ethnic groups and brought political stability and economic progress.

After Houphouët-Boigny died in 1993, the constitution called for the speaker of the National Assembly, Henri Konan Bédié, to take over as president for the rest of the term. In 1995, Bédié was elected president. In 1999, however, military officers led by General Robert Gueï ousted Bédié and set up a transitional government.

In 2000, voters approved a new constitution designed to return the country to civilian rule. A presidential election was held, but when it appeared that candidate Laurent Gbagbo would defeat General Gueï, Gueï shut down the election commission and declared himself the winner. Mass protests forced Gueï to leave office, and Gbagbo was sworn in as president. Former Prime Minister Alassane Ouattara was barred from running in both the elections because of questions about his nationality. In late 2000 and early 2001, clashes took place between Ouattara's and Gbagbo's supporters.

In 2002, a group of soldiers tried to overthrow the Gbagbo government. The coup attempt failed, but the rebels seized most of northern Côte d'Ivoire. Later that year, rebel uprisings began in western Côte d'Ivoire. The three rebel groups, the Ivorian government, and opposition parties agreed to form a power-sharing government in 2003. However, several cease-fire agreements failed to end the conflict, and disagreements about voting eligibility caused elections to be postponed repeatedly. Elections were finally held in 2010. Gbagbo and Ouattara both claimed victory, and violence erupted. Many countries and international organizations endorsed Ouattara's victory. In 2011, Gbagbo was arrested, and Ouattara was sworn in as president.

# PEOPLE AND ECONOMY

Almost all the people of Côte d'Ivoire belong to four major African ethnic groups. The Akan live in the southeast, the Kru in the southwest, the Voltaic in the north and northeast, and the Mandé in the west-central and northwest regions.

These major groups are actually made up of more than 60 smaller ethnic groups. French is the nation's official language, but many languages are spoken by the various ethnic groups. The Jula language, which is used in trade, is the country's most widely spoken language.

Since the mid-1900's, a large number of immigrants—mainly from Burkina Faso, Mali, and Guinea—have moved to Côte d'Ivoire. The population also includes many people of Lebanese or French descent.

Most Ivorians used to practice ancient local religions, but now about two-fifths of the people are Muslims, and about one third are Christians. The largest Christian church in Africa and one of the largest in the world, Our Lady of Peace, is located in the city of Yamoussoukro. The church was dedicated in 1990.

Drummers beat out rhythms for Ivorian dancers. Most Ivorians practice traditional religions, and their dances often have religious significance.

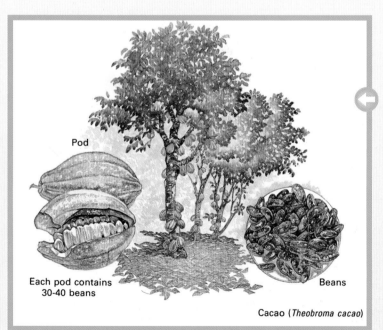

Pod

Each pod contains 30-40 beans

Beans

Cacao (*Theobroma cacao*)

Cacao trees are the source of the beans used to make cocoa and cocoa butter. The light brown to purple cacao beans develop inside melon-like fruits that grow from the tree's trunk and branches.

Workers handle sacks of coffee beans at Abidjan, one of western Africa's major ports. Coffee and cacao have long been Côte d'Ivoire's chief cash crops. The French grew coffee and cacao trees in their colony after World War I.

A member of one of the Akan ethnic groups of southeastern Côte d'Ivoire is marked by a pattern of traditional facial scars.

## Rural and urban life

About half of the Ivorian people are farmers who live in small villages. Each village family has its own *compound,* or group of homes that house members of an *extended family.* An extended family is one that includes such relatives as parents, married children and their offspring, aunts, uncles, and cousins. The homes usually have mud walls and roofs made of thatch or metal.

Agriculture is the country's chief source of income. Côte d'Ivoire leads the world in the production of *cacao beans,* which are used to make cocoa. Cacao, coffee, and palm oil are the country's chief exports. Other exports include bananas, cotton, petroleum products, pineapples, and rubber. Ivorian farmers also grow cassava, corn, rice, and yams and raise cattle, sheep, and goats.

Côte d'Ivoire is one of western Africa's most industrialized countries. A number of industries and manufacturing plants process the country's raw products. These operations include petroleum refining; the processing of palm oil, pineapples, sugar, timber, and tuna; and the production of textiles.

Since the mid-1900's, many young rural Ivorians have moved to the cities to find work, and now about half of the people live in urban areas. In the cities, sharp contrasts exist between middle- and upper-income households and poor households in housing, health, employment, and education.

## Transportation and education

Côte d'Ivoire has one of Africa's best transportation systems. A railroad operates between Abidjan and Ouagadougou, Burkina Faso. It can handle large numbers of passengers and hundreds of thousands of tons of freight every year.

Côte d'Ivoire also has several harbors, the two largest at Abidjan and San Pédro. Abidjan is one of the busiest seaports in western Africa, as well as an industrial and commercial center. In addition, an international airport also operates at Abidjan, and a number of smaller airports serve other parts of the country.

About half of Ivorians 15 years of age and older can read and write. The University of Cocody in Abidjan is one of the nation's chief institutions of higher learning. Abidjan is also known for its culture and its music industry.

# CROATIA

The republic of Croatia lies in the Balkan Peninsula in southeastern Europe. Formerly one of the six republics of Yugoslavia, Croatia broke away from the federation and declared its independence in 1991.

Croatia's varied and beautiful landscape includes the Adriatic coastlands along its western border, the scenic slopes of the Dinaric Alps running through its interior, and the fertile Pannonian Plains bordering Hungary to the north. Croatia's two principal rivers are the Drava and the Sava.

Zagreb, Croatia's capital, is the chief manufacturing center in this highly industrialized country. Croatia produces cement, chemicals, food products, petroleum, ships, steel, and textiles.

Most of Croatia's people belong to a Slavic group called Croats and speak Croatian. Most Croats follow the Roman Catholic religion. Present-day Croats are descended from Slavic tribes that began to settle in what is now Croatia during the A.D. 600's. Throughout the 900's and 1000's, Croatia flourished as an independent kingdom.

After about 1100, Croatia formed a political association with Hungary. In 1526, the Ottoman Empire, based in what is now Turkey, defeated Hungary. The Ottomans took control of most of Croatia, and the remainder came under the rule of the Habsburg family, which ruled Austria. A treaty in 1699 turned the Ottoman area of Croatia over to the Habsburgs. In 1918, Croatia united with neighboring territories to form a new country—the Kingdom of the Serbs, Croats, and Slovenes. In 1929, the kingdom's name was changed to Yugoslavia.

After World War II (1939-1945), Yugoslavia became a Communist country. It consisted of six republics—Bosnia-Herzegovina, Croatia, Macedonia, Montenegro, Serbia, and Slovenia. Eventually, many Croats came to resent the dominant role of Serbia in Yugoslavia's government.

In 1990, the Communist Party gave up its monopoly on power in Yugoslavia. Non-Communists

## FACTS

| | |
|---|---|
| ● Official name: | Republika Hrvatska (Republic of Croatia) |
| ● Capital: | Zagreb |
| ● Terrain: | Geographically diverse; flat plains along Hungarian border, low mountains and highlands near Adriatic coastline and islands |
| ● Area: | 21,851 mi² (56,594 km²) |
| ● Climate: | Mediterranean and continental; continental climate predominant with hot summers and cold winters; mild winters, dry summers along coast |
| ● Main rivers: | Drava, Sava |
| ● Highest elevation: | Mount Dinara, 6,007 ft (1,831 m) |
| ● Lowest elevation: | Adriatic Sea, sea level |
| ● Form of government: | Presidential/parliamentary democracy |
| ● Head of state: | President |
| ● Head of government: | Prime minister |
| ● Administrative areas: | 20 zupanije (counties), 1 grad (city) |
| ● Legislature: | Sabor (Assembly) with 152 members serving four-year terms |
| ● Court system: | Supreme Court, Constitutional Court |
| ● Armed forces: | 18,600 troops |
| ● National holiday: | Independence Day - October 8 (1991) |
| ● Estimated 2010 population: | 4,440,000 |
| ● Population density: | 3 persons per mi² (78 per km²) |
| ● Population distribution: | 57% urban, 43% rural |
| ● Life expectancy in years: | Male, 72; female, 79 |
| ● Doctors per 1,000 people: | 2.5 |
| ● Birth rate per 1,000: | 9 |
| ● Death rate per 1,000: | 12 |
| ● Infant mortality: | 6 deaths per 1,000 live births |
| ● Age structure: | 0-14: 16%; 15-64: 67%; 65 and over: 17% |
| ● Internet users per 100 people: | 51 |
| ● Internet code: | .hr |
| ● Languages spoken: | Croatian, Serbian |
| ● Religions: | Roman Catholic 87.8%, Orthodox 4.4%, Muslim 1.3%, other 6.5% |
| ● Currency: | Croatian Kuna |
| ● Gross domestic product (GDP) in 2008: | $69.33 billion U.S. |
| ● Real annual growth rate (2008): | 4.8% |
| ● GDP per capita (2008): | $15,786 U.S. |
| ● Goods exported: | Chemicals, food, machinery, petroleum, vehicles |
| ● Goods imported: | Chemicals, food, machinery, petroleum, transportation and electrical equipment |
| ● Trading partners: | Austria, Bosnia and Herzegovina, Germany, Italy, Russia |

doned Communism and began to establish *free enterprise*, a system in which businesses operate without government control. Croatia and Slovenia wanted to change Yugoslavia into a union of independent states. Serbia and Montenegro wanted a centralized state. In 1991, Croatia and Slovenia each declared independence.

War broke out between the Croatian and Serbian ethnic groups living in Croatia. The Yugoslav military fought on the side of the Serbs. Within a few months, Serbian forces had taken over about 30 percent of Croatia's land. A cease-fire agreement in January 1992 ended most of the fighting.

In 1995, Croatian forces began taking back the land seized by the Serbs. The Croatian government and the leaders of the Croatian Serbs made peace later that year. Croatia and Yugoslavia agreed to normalize relations. In 1998, the remaining land that had been seized by Croatian Serbs was returned to Croatia.

Through the 2000's, each of Croatia's democratically elected governments has worked to qualify for membership in the European Union. In 2009, Croatia joined NATO.

**Workers tend the grapevines in a vineyard in the coastal region of Dalmatia, now part of Croatia. Citrus fruits, figs, olives, and tobacco also flourish in this region.**

**Formerly one of the federal republics of Yugoslavia, Croatia declared its independence in 1991. Zagreb, its capital and largest city, is also a major trade, industrial, and cultural center.**

**Croatian women stop for a chat on Susak, a remote island off Croatia's Adriatic coast. Susak, the only place where the ancient Croatian dialect is still spoken, has no electricity or automobiles.**

# CUBA

Christopher Columbus claimed Cuba, an island country in the Caribbean Sea, for Spain in 1492. In the 1500's, Spanish settlers established plantations on the island and forced the native Indians to work on them. Many Indians died of diseases or harsh treatment. As the native population declined, the Spaniards began to bring Africans to work as slaves.

## Struggle for independence

In 1812, a group of slaves headed by José Antonio Aponte planned a revolt. The Spaniards discovered the plan and hanged Aponte, but various other groups continued to plot against Spain.

Cuba's struggle to end Spanish rule led to the Ten Years' War, which began in 1868. The war ended with a treaty that promised the abolition of slavery. Slavery ended in 1886, but many Cubans still wanted independence. A revolution broke out in 1895, led by José Martí. By 1898, Spain had lost control of all of Cuba except the major coastal cities.

United States President William McKinley told Spain to either change its policy in Cuba or give up the island. McKinley sent the U.S. battleship *Maine* to Havana to protect U.S. citizens in Cuba. When the *Maine* mysteriously blew up, the United States blamed the Spanish and declared war on Spain.

Spain lost the war and rights to Cuba in 1898. The United States then set up a military government on the island, which angered many Cubans and Americans alike.

## U.S. involvement

Under strong pressure from the Cuban people, the United States decided to let Cubans govern themselves, and Cuba adopted a constitution in 1901. However, the United States insisted that the document allow the United States to intervene in Cuban affairs. The United States was also allowed to buy or lease land for naval bases in Cuba. In 1903, the United States leased Guantánamo Bay and built a large naval base there, which remains today.

A huge portrait of Fidel Castro (far right) and his fellow revolutionary leader Che Guevara is displayed in Havana's Plaza de la Revolución.

The United States removed its troops in 1902, but it sent armed forces into Cuba on three subsequent occasions when revolts threatened the Cuban government. Many businesses, factories, and farms in Cuba were owned by U.S. companies, and the U.S. troops were sent to protect this property.

In 1933, an army sergeant named Fulgencio Batista helped overthrow the government. Eventually, he removed the president of the new government because the United States did not support him. Batista felt his best hope for power lay in winning U.S. support for himself.

From 1934 to 1940, Batista ruled Cuba as dictator through presidents who served in name only. In 1940, Cubans elected him president. The Constitution prevented him from being reelected in 1944, but in 1952 Batista again overthrew the government.

Cuba prospered under Batista. However, he was a dictator, and most Cubans continued to live in poverty. In 1953, a young lawyer named Fidel Castro started a revolution against Batista. Fidel, his brother Raúl, and many of their followers were imprisoned. On their release, the Castro brothers formed a revolutionary group.

## Revolution and Communist rule

The Castro brothers' forces launched guerrilla attacks on the government. On Jan. 1, 1959, Batista was forced to flee the country, and Fidel Castro became premier of Cuba. Many former political officials and army officers were tried and executed. Many other Cubans fled the country.

Old Havana, the oldest part of the capital, has many graceful and ornate buildings from the 1700's and 1800's.

Cuban drivers manage to keep American cars from the 1940's and 1950's running despite a lack of parts. Few new vehicles have entered Cuba in decades because of the general poverty and the U.S. embargo.

Castro's government established the first Communist state in the Western Hemisphere. In 1959 and 1960, it seized U.S.-owned farms and businesses. Castro turned to the Soviet Union for support. In 1961, the United States ended diplomatic relations with Cuba.

In April 1961, Cuban exiles sponsored by the United States Central Intelligence Agency (CIA) invaded Cuba at the Bay of Pigs. Castro's forces crushed the invasion.

In 1962, the U.S. began a naval blockade of Cuba when the Soviet Union began shipping nuclear missiles and materials to build launch sites on the island. Cuba had asked for such aid from the Soviet Union because it feared an attack by the United States. The Cuban missile crisis brought the world close to nuclear war. The tension ended when the Soviets removed the weapons—over Castro's protest.

Fidel Castro remained in firm control of Cuba for nearly 50 years, first as prime minister, and later as president. Raúl Castro succeeded Fidel as Cuba's president in 2008.

The island of Cuba is one of the most beautiful places in the group of Caribbean islands called the Greater Antilles. Cubans call it the *Pearl of the Antilles*. It lies about 90 miles (140 kilometers) south of Florida. The island of Cuba and more than 1,600 smaller islands make up the Republic of Cuba.

## The land

Cuba has a varied and magnificent landscape. Towering mountains and rolling hills cover about a fourth of the island. The rest of Cuba consists mainly of gently sloping land, grassy plains, and wide, fertile valleys. Its coastline is marked with deep bays, sandy beaches, and colorful coral reefs.

Cuba has three mountainous regions—the Sierra de los Órganos and Sierra del Rosario ranges in the northwest, the ranges of the Sierra de Escambray region in central Cuba, and several ranges, including the heavily forested Sierra Maestra, in the southeast.

The rich cropland and pastureland between the mountain ranges consist mainly of red clay. Parts of the coastline are bordered by lowlands and swamps.

Cuba's 2,100-mile (3,380-kilometer) coastline is indented with about 200 harbors. The larger harbors have narrow entrances to the sea, so the ships they shelter are protected against wind and waves. The most important harbors include Havana and Nuevitas on the north coast and Cienfuegos, Guantánamo, and Santiago de Cuba on the south coast.

Coral islands and reefs lie off the coast. The largest of the more than 1,600 islands that surround the Cuban mainland is Isla de la Juventud (the Isle of Youth).

Lying within the northern tropics, Cuba has a semitropical climate. Cool ocean breezes from the northeast in the summer and warm breezes from the southeast in the winter give the island a mild climate throughout the year. However, Cuba lies within a hurricane area. These violent windstorms frequently hit the island, especially its eastern and western tips, during the hurricane season from June to November.

## FACTS

| | |
|---|---|
| **Official name:** | Republica de Cuba (Republic of Cuba) |
| **Capital:** | Havana |
| **Terrain:** | Mostly flat to rolling plains, with rugged hills and mountains in the southeast |
| **Area:** | 42,427 mi² (109,886 km²) |
| **Climate:** | Tropical; moderated by trade winds; dry season (November to April); rainy season (May to October) |
| **Main rivers:** | Cauto, Salado, San Pedro, Caunao, Zaza |
| **Highest elevation:** | Pico Turquino, 6,542 ft (1,994 m) |
| **Lowest elevation:** | Caribbean Sea, sea level |
| **Form of government:** | Socialist state |
| **Head of state:** | President of the Council of State and the Council of Ministers |
| **Head of government:** | President of the Council of State and the Council of Ministers |
| **Administrative areas:** | 14 provincias (provinces), 1 municipio especial (special municipality) |
| **Legislature:** | Asemblea Nacional del Poder Popular (National Assembly of People's Power) with 614 members serving five-year terms |
| **Court system:** | Tribunal Supremo Popular (People's Supreme Court) |
| **Armed forces:** | 49,000 troops |
| **National holiday:** | Triumph of the Revolution - January 1 (1959) Anniversary of Fidel Castro's attack on the Moncada Army Barracks - July 26 (1953) |
| **Estimated 2010 population:** | 11,265,000 |
| **Population density:** | 266 persons per mi² (103 per km²) |
| **Population distribution:** | 76% urban, 24% rural |
| **Life expectancy in years:** | Male, 75; female, 79 |
| **Doctors per 1,000 people:** | 5.9 |
| **Birth rate per 1,000:** | 10 |
| **Death rate per 1,000:** | 7 |
| **Infant mortality:** | 5 deaths per 1,000 live births |
| **Age structure:** | 0-14: 18%; 15-64: 70%; 65 and over: 12% |
| **Internet users per 100 people:** | 13 |
| **Internet code:** | .cu |
| **Language spoken:** | Spanish |
| **Religions:** | Roman Catholic, Protestant, Jewish, Santeria |
| **Currency:** | Cuban peso, Cuban convertible peso (chavito) |
| **Gross domestic product (GDP) in 2008:** | $55.18 billion U.S. |
| **Real annual growth rate (2008):** | 4.3% |
| **GDP per capita (2008):** | $4,853 U.S. |
| **Goods exported:** | Citrus fruits, coffee, fish and shellfish, medical products, nickel, rum, sugar, tobacco products |
| **Goods imported:** | Machinery, petroleum, wheat and other food products |
| **Trading partners:** | Canada, China, Spain, Netherlands, Venezuela |

UNITED STATES

Straits of Florida

BAHAMAS

Gulf of
Mexico

Great Bahama Bank

North

CUIDAD DE
LA HABANA
Havana
San Miguel del Padrón
Bahía de Cárdenas
Matanzas
Cárdenas
Corralillo
Archipiélago de Sabana
Bahía Honda
Mariel
San José de las Lajas
Jovellanos
Perico
Sagua
la Grande
Old Bahama Channel
La Palma
Sierra del
Rosario
Güines
LA
HABANA
Colón
Santo
Domingo
Cifuentes
Bahía de
Buena Vista
Archipiélago de Camagüey
PINAR DEL RÍO
Artemisa
San Cristóbal
Consolación del Sur
MATANZAS
Jagüey Grande
Santa Clara
VILLA CLARA
Camajuaní
Cayo
Coco
Cayo
Romano
Sierra de los
Órganos
Pinar del Río
Golfo de
Batabanó
Peninsula
de Zapata
CIENFUEGOS
Ranchuelo
Placetas
Yaguajay
Guane
San Juan
y Martínez
Cienfuegos
Manicaragua
Cabaiguán
Chambas
Morón
Cabo
n Antonio
Ciudad Sandino
Río San Diego
Archipiélago de los Canarreos
Cumanayagua
Sierra de
Trinidad
SANCTI SPÍRITUS
CIEGO DE
ÁVILA
Esmeralda
Cabo
Corrientes
Nueva Gerona
Bahía de Cochinos
(Bay of Pigs)
Trinidad
Sancti
Spíritus
Ciego
de Ávila
Bahía de Nuevitas
North Atlantic
Ocean
Isla de la
Juventud
La Fé
ISLA DE LA JUVENTUD
(MUNICIPIO)
Cayo Largo
Río Marati
Florida
Minas
Nuevitas
Golfo de
Ana María
Camagüey
CAMAGÜEY
Manatí
Gibara
Cabo Lucrecia
Jardines de la Reina
Vertientes
San Pedro
Puerto Padre
LAS TUNAS
Jesús Menéndez
Banes
Bahía de Nipe
Santa Cruz
del Sur
Guáimaro
Las Tunas
Holguín
Antilla
Moa
Punta Guarico
Amancio
Bágoanos
Mayarí
Sagua de Tánamo
Guayabal
Golfo de
Guacanayabo
GRANMA
HOLGUÍN
Cabo
Maisí
Manzanillo
Bayamo
Mella
Palma Soriano
GUANTÁNAMO
Baracoa
Niquero
Yara
Contramaestre
Jiguaní
San Luis
Guantánamo
Caimanera
Sierra Maestra
SANTIAGO DE CUBA
Cabo Cruz
Pilón
Pico Turquino
6,542 ft
(1,994 m)
Santiago
de Cuba
Bahía de
Guantánamo
UNITED STATES
NAVAL BASE

Caribbean Sea

CUBA

Cuba is the westernmost
island in the Greater
Antilles, about 90 miles
(140 kilometers) south
of Florida.

Cayman Islands
(U.K.)

HAITI

JAMAICA

Cuba's landscape includes gentle
slopes and rolling plains against a
backdrop of green hills.

## The government

According to the Constitution of Cuba, the country is a socialist state. It is controlled by the Communist Party of Cuba, the
only party allowed. Membership in the party was restricted
until the 1990's, when the party began trying to broaden its
membership.

The government of Cuba is highly centralized, and political
and economic freedom is limited. The legislature is called the
National Assembly of People's Power. Members are elected to
five-year terms. It meets in only two sessions each year. The
National Assembly in turn elects the members of the Council
of State. The president of the council serves as head of state
and head of government, and is the most powerful government
official. The president, with the approval of the Assembly, appoints a Council of Ministers, which
serves as a cabinet, enforces laws, and directs government agencies.

Fidel Castro established firm control over the Cuban
government after leading a revolutionary take-over in
1959. He ruled as a dictator, first in the office of prime
minister and later as president. In 2006, Castro became
ill and temporarily gave control of the government to his
brother Raúl. In 2008, Fidel gave up the leadership of
Cuba, and the National Assembly elected Raúl to succeed
him. Raúl's government made several economic reforms,
including allowing farmers greater control of land use
and allowing people to buy and sell their homes. In 2009
and 2011, United States President Barack Obama eased
several restrictions that previous U.S. presidents had
placed on Cuba.

# PEOPLE AND ECONOMY

More than 11 million people live in Cuba, and approximately three-fourths of them live in the cities and towns. Havana, the capital and largest city, has a population of more than 2 million. Many of the rural people are poor. Few people live on the smaller islands off the coast.

## Ancestry

Most Cubans are descendants of people who came to the island from Spain or Africa. About 30 to 40 percent of Cuba's population is descended from people who came from Spain. About 10 to 20 percent are descended from Africans. Most of the remaining people have ancestors from both places. The island also has a small percentage of people of Chinese descent. Spanish is the official language.

Salsa band members play popular Cuban music. The government supports the arts.

## The economy

From 1961 to the early 1990's, government planning dominated key economic decisions in Cuba. A U.S. trade embargo imposed in 1960 contributed to the stagnation of Cuba's economy during this period. As a result, Cuba relied heavily on aid from, and trade with, the Soviet Union and other Communist nations.

Communism collapsed in Eastern Europe during the late 1980's, and the Soviet Union broke apart in 1991. As a result, Cuba lost much of its trade and suffered an economic crisis. The government loosened its hold on the economy. It allowed more foreign investment, mainly in the tourist industry, and some private business ownership. Many Cubans opened small businesses, including restaurants.

Cuba's trading patterns also changed. Today, Cuba's chief trading partners are Brazil, Canada, China, Germany, Italy, Mexico, Russia, Spain, and Venezuela.

During the 1960's, about three-fourths of Cuba's land had come under state control, and most farms were *state farms,* owned and run by the government. Other farms were organized as *farm cooperatives,* owned jointly by the government and groups of farmers. Some small farms remained under the control of individual owners, producing chiefly coffee and tobacco. In all cases, farmers were required to sell their products to the state at prices set by the government.

In the early 1990's, the government made more state farms into cooperatives. After meeting government-set production quotas, farmers could sell the rest of what they produced on the open market. Soon, small farmers' markets sprang up across the island to sell a variety of products directly to the public. Cuba's armed forces also manage farms and are a major food producer.

Sugar cane, coffee, and tobacco are the main crops. Farmers also raise other food crops and farm animals.

Day care for young children is provided to make women available for Cuba's work force. Older children must go to school from ages 6 to 14. The government runs the schools, and education is free.

A Cuban boy helps lay bricks and mortar. The Cuban government has built much badly needed housing, but not enough to meet the demands of the population.

Farmworkers wear wide-brimmed straw hats to protect them from the sun. Cuba's sugar cane production has deceased since the late 1990's, but the country is still a major producer.

Sugar is Cuba's main manufactured product, though many sugar mills closed in the early 2000's because of a decrease in world sugar prices. Havana is famous for factories that product fine hand-rolled cigars. Cuba's other industrial products include agricultural machinery, cement, food and beverage products, petroleum, pharmaceuticals, and steel. Cuban mines produce cobalt, gold, iron ore, natural gas, nickel, stone, and a small amount of petroleum.

State-owned Cuban fishing fleets range over the Caribbean Sea and parts of the North Atlantic Ocean. The important catches include such shellfish as lobsters and shrimp. Caibarién, Cienfuegos, and Havana are important fishing ports.

Service industries produce services rather than manufactured goods or agricultural products. They include banking, education, and health care. One of Cuba's fastest-growing service industries is tourism. The Cuban government has formed joint ventures with foreign investors to build new hotels and to restore old ones. Tourists have arrived in growing numbers, mainly from Canada and Europe.

# DAILY LIFE

Life in Cuba today is very much influenced by the socialist government. Before the Castro revolution in 1959, the Cuban government used much of the nation's economic resources to help make Havana a luxurious and popular tourist center. But little was done to improve the lives of the Cuban people, particularly those in rural areas.

Castro's government turned attention from Havana to the countryside. One of the goals of the socialist government is economic and social equality, and the government has made great progress toward this goal. Today, Cuba spends large sums of money on housing and food for rural people and provides free education and medical care for all its people.

The majority of Cuban people live in urban areas. Havana is Cuba's capital, largest city, and commercial and cultural center. Most people in the cities have jobs in government agencies or government-owned businesses and factories, but some run small private businesses. Cuba's cities have a serious housing shortage. Many buildings, neglected for several decades, are in need of repair.

Most of the people in Cuba's rural areas work on farms. Some rural people live in traditional *bohíos*, which are thatch-roofed dwellings with dirt floors.

Many kinds of food are scarce in Cuba. To help make sure that even the poorest have enough to eat, the government created a system of food rationing in the 1960's. The system is designed to provide all households with minimum quantities of rice, beans, meat, chicken, eggs, sugar, milk, and coffee.

Cuban law requires children to go to school from ages 6 to 14. The government controls the schools, and education is free.

The government also has set up various adult education programs. During the early 1960's,

A strong sea wall protects Havana from the heavy swells of the Atlantic Ocean. More than 2 million people live in the city, once one of the world's luxury tourist centers.

students were recruited to teach illiterate Cubans how to read and write. Later, other education projects were established. As a result, nearly all Cuban adults are literate today. Cuba has dozens of universities and university-level educational institutions.

About half of Cuba's roads are paved. The Central Highway extends between Pinar del Río, in the northwest, and Santiago de Cuba, in the southeast. However, gasoline rationing, problems in obtaining spare parts, and a scarcity of new vehicles often have made automobile transportation difficult. Railroads cross the island, but more people ride buses than trains. José Martí International Airport is the country's largest airport.

Cuba's principal newspaper is *Granma,* published by the Cuban Communist Party. The government or the Communist Party controls all newspapers, most magazines, and television and radio broadcasts. However, Florida is only 90 miles (140 kilometers) away, and the Cuban people can pick up radio and TV broadcasts from the United States. Radio Martí, a U.S. government radio station that broadcasts programs to the Cuban people, began operating in Miami, Florida, in 1985.

Cannons still line the harbor of Havana, which was founded by the Spaniards in 1519. At one time, cannons were used to protect the town from raiding pirates.

Havana's treelined avenues and impressive buildings are a reminder of its past. Because the government has invested much more in rural areas, Havana now has a housing shortage.

About 40 percent of Cubans belong to the Roman Catholic Church. Many Cubans also believe in Santería, a religion that combines traditional African religious beliefs with Roman Catholic ceremonies. There are also some Protestant groups, and a small Jewish community is concentrated mostly in Havana. Until the 1990's, the People's Communist Party of Cuba, the only political party in the country, did not allow people who attended religious services to join it. In the 1990's, however, the party began to expand its membership and to include religious believers.

The Cuban government strongly supports the arts and sponsors free ballets, plays, and other cultural events. The work of the government-sponsored Cuban Institute of Cinematographic Art and Industry has made Cuba a center of the Latin American film industry. The Casa de las Américas (House of the Americas)—which supports the work of artists, writers, and musicians—is one of Latin America's most prestigious cultural institutions. Most Cubans enjoy singing and dancing, especially Cuban folk music and the country's traditional dances. Cubans are also enthusiastic sports fans who enjoy baseball, basketball, swimming, and track and field.

A female group performs folk dances and plays Cuba's traditional musical instruments, such as the conga drum.

# CYPRUS

Greek legend tells how Aphrodite, the goddess of love and beauty, sprang from the foam of the sea and was carried by the waves and sea breezes to the Isle of Cyprus. There, she was prepared and dressed to meet the assembly of gods.

This small island country, in the northeastern corner of the Mediterranean Sea, lies about 40 miles (64 kilometers) south of Turkey and 60 miles (97 kilometers) west of Syria. It is the third-largest island in the Mediterranean Sea, after Sicily and Sardinia. Geographically, the island is part of Asia, but its history and culture have long been linked to European civilization.

## Early history

The remains of a Neolithic settlement at Khirokitia are evidence that the island was settled as long ago as 6000 B.C. Greek settlers arrived around the 1100's B.C. Before the birth of Christ, Cyprus had been conquered by the Phoenicians, Assyrians, Egyptians, Persians, Macedonians, and Romans.

The island was part of the East Roman (Byzantine) Empire starting in 395. It was ruled by the Lusignans, a French family, from 1192 until 1489, when Venice took control. The Ottoman Empire, based in Turkey, conquered Cyprus in the 1570's. In 1878, the United Kingdom and the Ottoman Empire signed an agreement that allowed the British to administer the island. Cyprus became a crown colony of the United Kingdom in 1925.

## Independence and conflict

In the 1950's, Archbishop Makarios III led a Cypriot campaign for *enosis*—that is, union with Greece. After terrorist attacks by a Greek Cypriot secret organization called EOKA, the British declared a state of emergency on the island. In 1959, British, Cypriot, Greek, and Turkish leaders meeting in Zurich, Switzerland, agreed that Cyprus should be an independent state. Cyprus became independent on Aug. 16, 1960, and Archbishop Makarios became president of the new state.

## FACTS

| | |
|---|---|
| Official name: | Republic of Cyprus |
| Capital: | Nicosia |
| Terrain: | Central plain with mountains to north and south; scattered but significant plains along southern coast |
| Area: | 3,572 mi² (9,251 km²) |
| Climate: | Temperate, Mediterranean with hot, dry summers and cool winters |
| Main rivers: | Pedieos, Serakhis, Ezouza, Dhiarrizos, Kouris |
| Highest elevation: | Mount Olympus, 6,403 ft (1,952 m) |
| Lowest elevation: | Mediterranean Sea, sea level |
| Form of government: | Republic |
| Head of state: | President |
| Head of government: | President |
| Administrative areas: | 6 districts |
| Legislature: | Greek area: Vouli Antiprosopon (House of Representatives) with 80 members serving five-year terms<br>Turkish area: Cumhuriyet Meclisi (Assembly of the Republic) with 50 members serving five-year terms |
| Court system: | Supreme Court |
| Armed forces: | 10,000 troops |
| National holiday: | Greek area: Independence Day - October 1 (1960)<br>Turkish area: Independence Day - November 15 (1983) |
| Estimated 2010 population: | 812,000 |
| Population density: | 227 persons per mi² (88 per km²) |
| Population distribution: | 66% urban, 34% rural |
| Life expectancy in years: | Male, 75; female, 80 |
| Doctors per 1,000 people: | 33.3 |
| Birth rate per 1,000: | 13 |
| Death rate per 1,000: | 8 |
| Infant mortality: | 7 deaths per 1,000 live births |
| Age structure: | 0-14: 18%;<br>15-64: 71%;<br>65 and over: 11% |
| Internet users per 100 people: | 40 |
| Internet code: | .cy |
| Languages spoken: | Greek, Turkish, English |
| Religions: | Greek Orthodox 78%, Muslim 18%, other 4% |
| Currency: | Euro |
| Gross domestic product (GDP) in 2008: | $22.69 billion U.S. |
| Real annual growth rate (2008): | 3.6% |
| GDP per capita (2008): | $30,016 U.S. |
| Goods exported: | Greek area: food, machinery, pharmaceuticals, vehicles<br>Turkish area: citrus, dairy, potatoes, textiles |
| Goods imported: | Greek area: machinery, petroleum, pharmaceuticals, transportation equipment<br>Turkish area: food, fuel, minerals, vehicles |
| Trading partners: | Greek area: Germany, Greece, United Kingdom<br>Turkish area: Turkey |

In 1963, disagreements over changes in the Cypriot Constitution proposed by President Makarios led to renewed conflict between the Greek and Turkish Cypriots. The Turkish Cypriots feared that Makarios' changes would eliminate many of their rights, and fighting broke out.

Between 1967 and 1974, some progress was made toward agreement between the two sides. But in 1974, Greek officers, supported by military rulers in Greece, overthrew President Makarios. Then Turkey invaded the island, and new fighting erupted. Turkish forces captured large areas of the northeast, and thousands of Greek Cypriots fled to southwestern Cyprus. Thousands of Turkish Cypriots moved to the north.

A cease-fire in August 1974 ended the fighting, and Makarios returned as president. In 1975, Turkish Cypriot leaders declared the northeast part of the island an autonomous territory within Cyprus, and in 1983, they declared it independent. However, most countries did not recognize their declaration of independence.

Since 1974, representatives of the Greek and Turkish Cypriots, and of Greece and Turkey, have frequently met to discuss the continuing problem. In 1998, Cyprus began talks to join the European Union, and it became a

Cyprus is a scenic island of colorful folk traditions, beautiful beaches, and historic art and architecture.

member in 2004. In 2003, the Turkish Cypriot government lifted restrictions on travel between the northern and southern parts of Cyprus. In 2008, a major crossing between the Greek and Cypriot sections of Nicosia, the capital city, was opened.

Cyprus is an island nation in the northeastern Mediterranean Sea about 40 miles (64 kilometers) south of Turkey. A military barrier, called the Attila line, has divided the Greek-speaking south from the Turkish-held north since 1974.

# TOURISM

Scenic beauty and an average of 340 days of sunshine a year attract many visitors to the island paradise of Cyprus, making tourism an important part of the economy. The island boasts a wealth of ruins from ancient times, as well as magnificent Gothic art and architecture. Golden, sandy beaches attract sunbathers, while the snow on Mount Olympus invites skiing enthusiasts.

Political tension on Cyprus has slowed the island's economic progress, however. The division of the island into Greek and Turkish zones hurt the tourism trade during the 1970's, but tourism has been revived in the Greek area since then.

## Landscape

The mountains of the Troodos Massif rise in the southwest region of the island, while the Kyrenia range stretches along the northern coast. Extensive pine forests cover parts of the Troodos. The mountains also have rich deposits of asbestos and chromite.

Between the island's two mountain ranges lie the Mesaoria and Morphou plains. The fertile soil and mild climate of the plains yield two annual harvests of fruits and vegetables. Agricultural production makes a major contribution to the island's economy. Wheat and barley are grown in the eastern plain, while citrus fruits, olives, and almonds are grown in the west.

Extensive forests cover the rugged slopes of the Troodos Massif in southwestern Cyprus, where the island's highest peak, Mount Olympus, rises to 6,403 feet (1,952 meters).

## A mix of cultures

Tourists visiting Cyprus encounter a mix of cultural and ethnic groups, reflecting the island's long history of invasion. About three-fourths of Cypriots live in the Greek Cypriot section. The majority of Greek Cypriots belong to the Orthodox Church of Cyprus, one of the Eastern Orthodox Churches. Most Turkish Cypriots are Sunni Muslims. Many Eastern Europeans have immigrated to southern Cyprus. Since 1974, many Turks have settled in the north. The large number of Turkish settlers in the north has been a major issue in reunification negotiations. The island's population also includes small groups of Armenians, Maronites, and Roma.

Greek is the primary language of the south, and Turkish is the language of the north. The Cypriot dialects of both languages share many words. English is widely spoken by both communities.

Fishers at the south coastal port of Limassol prepare freshly caught octopus for market. Like Larnaca to the northeast, Limassol is a major port as well as a bustling center of industry and commerce.

Boats are often moored in the Turkish-occupied port of Kyrenia, a resort on the north-central coast.

## Historic sites

The romance and legend of Cyprus's past come vividly to life in Paphos, the island's most important town during Roman times. The picturesque harbor and the defensive wall of the old port at Paphos were built by Alexander the Great. Near the harbor, the mosaic pavements of the Villa of Dionysus are among the most beautiful in the Mediterranean area.

At the village of Kouklia (Palea Paphos) are the remains of the Sanctuary of Aphrodite, where pilgrims worshiped the goddess in ancient times. In the ancient city of Kourion, archaeologists have uncovered ruins dating from the 400's B.C. Fine pavement mosaics, baths, and a Temple of Apollo are all that survived after an earthquake destroyed the city.

A few miles east of Kourion stands the massive Kolossi Castle, built in 1210 by the Knights of St. John. From the castle's towers, visitors may enjoy an extraordinary view of the rolling plains between the Troodos Massif and the city of Limassol, where King Richard the Lion-Hearted of England, who conquered and then sold the island in 1191, married Berengaria of Navarre.

In the interior of the island, the capital city of Nicosia has an old section surrounded by a huge ring of Venetian walls. Inside the walls stands the Gothic St. John's Cathedral, seat of the Greek Orthodox Archbishop of Cyprus.

In the port of Famagusta is a fortress featuring Othello's Tower, said to have been the scene of Shakespeare's tragedy. Not far from Famagusta lies the site of the ancient city of Salamis. Once a shipping center for the island's busy copper trade, it was destroyed by an earthquake in the A.D. 300's. The ruins of a large theater, a Temple of Zeus, an aqueduct, and other structures still stand today.

On the hill of Throni, outside the capital city of Nicosia, soldiers guard the tomb of President Archbishop Makarios III. In the 1950's, Makarios led the Greek Cypriot campaign for independence from British rule.

The Czech Republic is a landlocked nation in the heart of central Europe, bordered by Poland to the north, Slovakia to the east, Austria to the south, and Germany to the west. Its scenic landscape ranges from rugged mountains and rolling hills to dense forests and fertile plains. Prague, the nation's capital and largest city, has been called the "City of a Hundred Spires" because of its many churches.

Two historic regions—Bohemia in the west and Moravia in the east—make up most of the republic. The northern part of the country also includes a small part of a region called Silesia, which extends into Poland.

## A new nation

The Czech Republic was formerly a part of Czechoslovakia, which was established in 1918 when the Czech and Slovak people united to form a single nation. In mid-1992, Czech and Slovak leaders began discussions on whether to split the nation into two republics. Czechoslovakia became two separate nations—the Czech Republic and Slovakia—on Jan. 1, 1993, as the result of a peaceful "divorce" engineered by the leaders.

The first known inhabitants of what is now the Czech Republic were the Boii, a Celtic tribe that lived in Bohemia in about the 400's B.C. By about A.D. 500, Slavic tribes, including the ancestors of the Czechs, settled in the region. In the 800's, several Slavic tribes joined together to form the Greater Moravian Empire, which covered much of central Europe until it fell around 900.

Bohemia then expanded. It came under the protection of the Holy Roman Empire, a German-based empire that included much of central Europe. Moravia became part of the territory ruled by the Duke of Bohemia in the 900's. In the 1100's to 1200's, Bohemia became a semi-independent kingdom within the Holy Roman Empire. In 1526, the Habsburgs, who ruled Austria, also began ruling Bohemia.

In 1620, Bohemia lost its self-governing powers. It was divided into three provinces— Bohemia, Moravia, and Silesia—ruled by the Habsburgs. The Habsburgs forced the Czechs to adopt the German language and culture.

144

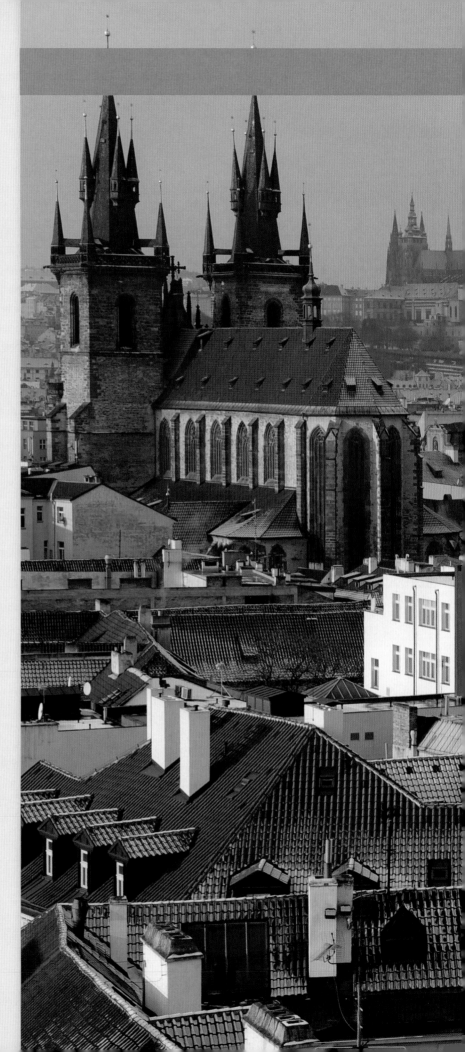

# CZECH REPUBLIC

In 1867, Austria and Hungary formed a monarchy called Austria-Hungary, which lasted until the end of World War I (1914-1918). During this period, Czech and Slovak leaders worked together to form an independent nation. After the collapse of Austria-Hungary in 1918, Czechoslovakia was proclaimed an independent country. In 1948, Czechoslovakia became a Communist state.

Although the Czechs and Slovaks were united under a single flag, the two groups followed separate paths and maintained their own traditions. Since the Slovaks were mainly farming people with very little political experience, the Czechs took control of the new nation's economy and government.

Despite their cultural differences, the Czechs and Slovaks united to seek political reform in the late 1980's. Working together, the two groups helped bring about the fall of Communism and the return of democratic government to their country.

When Czechoslovakia was dissolved in 1993, the Czechs and Slovaks cooperated to divide the assets of the country. They completed the transition through peaceful negotiations.

## The Czech people

The Czech people enjoy one of the highest standards of living in Eastern Europe. Most of the people live in towns and cities, usually in apartment buildings. People in rural areas usually work in agriculture or travel to cities or nearby factories to work. Rural families often live in single-family homes.

About 81 percent of the people of the Czech Republic are Czechs, and about 13 percent are Moravians. Slovaks make up about 3 percent of the population. Small numbers of Germans, Hungarians, Poles, and Roma (sometimes called Gypsies) also live in the Czech Republic.

The republic's official language is Czech. Moravians speak a form of Czech that is slightly different from that spoken in Bohemia. Slovaks speak Slovak, which is closely related to Czech. Some members of other minority groups speak their own languages at home but generally also speak Czech.

145

# CZECH REPUBLIC TODAY

November 1989 marked the beginning of dramatic political changes for the Czech people. These changes were triggered by the decline of Communist power in Czechoslovakia and the trend toward greater democracy in Eastern European countries during the late 1980's.

On Nov. 17, 1989, about 50,000 people took part in antigovernment demonstrations in Prague—the largest public protest in Czechoslovakia in 20 years. The police broke up the demonstration, but the protests continued and increased in size. The continuing demonstrations were not opposed by police, and a new opposition group, known as Civic Forum, called for a return to democracy. Civic Forum was supported by the majority of the people, as well as by the Czechoslovak Socialist Party and the Czechoslovak People's Party. In early December, a new federal government was announced, but it was denounced by the Civic Forum because most of its ministers had been members of the previous administration and it included only five non-Communists.

After a demonstration of about 200,000 people in Prague, a new, interim federal government was formed. It consisted of a non-Communist majority that also included seven supporters of Civic Forum. In addition, the Federal Assembly voted to change the Constitution, deleting the sections that guaranteed the Communist Party's leading role.

Gustáv Husák, leader of the Communist Party, resigned as president. At the end of December, he was replaced by playwright Václav Havel, leader of the Civic Forum. During this "velvet revolution," the word *Socialist* was removed from the nation's official name.

Under the new government, such civil liberties as freedom of religion, speech, and the press were restored. In 1992, the government also moved to quickly establish a free market economy. That same

## FACTS

| | |
|---|---|
| Official name: | Ceska Republika (Czech Republic) |
| Capital: | Prague |
| Terrain: | The west consists of rolling plains, hills, and plateaus surrounded by low mountains; the east consists of very hilly country |
| Area: | 30,450 mi$^2$ (78,866 km$^2$) |
| Climate: | Temperate; cool summers; cold, cloudy, humid winters |
| Main rivers: | Elbe, Morava, Vltava |
| Highest elevation: | Sněžka, 5,256 ft (1,602 m) |
| Lowest elevation: | Elbe River, 377 ft (115 m) |
| Form of government: | Parliamentary democracy |
| Head of state: | President |
| Head of government: | Prime minister |
| Administrative areas: | 13 kraje (regions), 1 hlavni mesto (capital city) |
| Legislature: | Parlament (Parliament) consisting of the Senat (Senate) with 81 members serving six-year terms and the Poslanecka Snemovna (Chamber of Deputies) with 200 members serving four-year terms |
| Court system: | Supreme Court, Constitutional Court |
| Armed forces: | 24,100 troops |
| National holiday: | Czech Founding Day - October 26 (1918) |
| Estimated 2010 population: | 10,202,000 |
| Population density: | 335 persons per mi$^2$ (129 per km$^2$) |
| Population distribution: | 74% urban, 26% rural |
| Life expectancy in years: | Male, 74; female, 80 |
| Doctors per 1,000 people: | 3.6 |
| Birth rate per 1,000: | 11 |
| Death rate per 1,000: | 10 |
| Infant mortality: | 3 deaths per 1,000 live births |
| Age structure: | 0-14: 14%; 15-64: 71%; 65 and over: 15% |
| Internet users per 100 people: | 48 |
| Internet code: | .cz |
| Languages spoken: | Czech, Slovak |
| Religions: | Roman Catholic 26.8%, Protestant 2.1%, unaffiliated 59%, other 12.1% |
| Currency: | Koruna |
| Gross domestic product (GDP) in 2008: | $216.92 billion U.S. |
| Real annual growth rate (2008): | 3.9% |
| GDP per capita (2008): | $21,256 U.S. |
| Goods exported: | Coal, iron and steel, machinery, transportation equipment |
| Goods imported: | Iron ore, machinery, natural gas, petroleum, pharmaceuticals, transportation equipment |
| Trading partners: | Austria, France, Germany, Italy, Poland, Russia, Slovakia |

year, Czech and Slovak leaders began discussing whether to divide Czechoslovakia into two nations, though a majority of the people opposed it. Havel resigned his post.

On Jan. 1, 1993, Czechoslovakia broke up into the Czech Republic and Slovakia. The Czech legislature elected Havel president. The Czech Republic continued to work toward a free enterprise economy. Many new businesses were established, especially in retail and other service industries. Today, about 55 percent of the labor force works in service industries.

Havel was elected to a second term in 1998. In 1999, the Czech Republic joined the North Atlantic Treaty Organization (NATO), a military alliance of Western nations. Václav Klaus, a co-founder of the Civic Democratic Party (CDP), was elected president in 2003. The country joined the European Union (EU) in 2004.

In parliamentary elections held in 2006, the CDP, led by Mirek Topolánek, won half the seats in the larger house of the legislature. Topolánek formed a coalition government (one made up of several political parties). The government lost a vote of confidence in 2009, and Topolánek resigned. An interim government was formed with Jan Fischer as prime minister. After parliamentary elections in 2010, Petr Necas of the CDP became prime minister.

The resort town of Mariánské Lázně in western Bohemia enjoyed great renown in the days of the Austro-Hungarian Empire, when it was known as Marienbad. Many people still come to bathe in its mineral springs or drink the waters—an activity that is considered healthful. Mariánské Lázně lies in the Bohemian Mountains, a region of large coal and uranium ore deposits and dense forests.

The Czech Republic came into existence on Jan. 1, 1993, when Czechoslovakia was officially dissolved.

# HISTORY

The ancestors of the Czech people settled in what is now the Czech Republic by about A.D. 500. During the 800's, these Slavic tribes banded together to form the Greater Moravian Empire. Beginning in the late 800's, the Magyars (Hungarians) invaded the Greater Moravian Empire, and in time they destroyed it.

### The rise of Bohemia

Bohemia then began to expand. In the mid-900's, Bohemia's ruling family also acquired Moravia. By the early 1200's, Bohemia was a semi-independent kingdom within the Germany-based Holy Roman Empire. During the 1200's, many German craft workers and merchants settled in Bohemian towns.

Bohemia reached its cultural and political peak under Charles IV, who ruled from 1346 to 1378. Charles was crowned Holy Roman Emperor in 1347, and Prague—the city in which he was born—became the empire's leading city. In 1415, the execution of the religious reformer John Hus triggered a series of religious wars in Bohemia. The Czech nobles, who had become supporters of church reform, grew increasingly powerful.

In 1526, Bohemia came under the control of the Austrian Habsburgs, who were Roman Catholic. Meanwhile, the Reformation gave birth to Protestantism in Europe, and in 1618, a group of Czech Protestant nobles revolted against Habsburg rule. They elected a Protestant king of Bohemia. The revolt touched off the Thirty Years' War.

The Bohemians were defeated by the Habsburgs in 1620, and Bohemia lost most of its self-governing powers. Bohemia and Moravia began to industrialize in the late 1700's, and many Czech peasants moved to urban areas to work in factory jobs. By the mid-1800's, the

An etching entitled *The Defenestration of Prague* illustrates Czech Protestants throwing Roman Catholic officials from a window of Prague Castle in 1618.

A painted ceiling depicting the great thinkers of ancient times decorates the Philosophical Hall of the Strahov Library near Prague Castle. This historic building on Hradčany (Castle Hill) houses many of the nation's artistic treasures.

Czechs had replaced the Germans as the largest population group in the cities and towns of Bohemia and Moravia. Nationalistic feelings began to develop among the Czechs, and several revolutionary leaders called for Czech self-government.

During World War I (1914-1918), opposition activity increased in Bohemia. Although the Habsburg government arrested many revolutionary leaders, two leading Czech nationalists—Tomáš G. Masaryk and Eduard Beneš—fled to Paris, where they formed the Czechoslovak National Council to organize foreign support for an independent state made up of Czechs and Slovaks.

### Unity of Czechs and Slovaks

When Austria-Hungary collapsed after World War I, Czechoslovakia became an independent democratic republic consisting of Bohemia, Moravia, and Slovakia. In 1919, Ruthenia, a region east of Slovakia, became part of Czechoslovakia.

Tension between minority groups weakened the new nation. In the 1930's, German dictator Adolf Hitler encouraged the Sudeten Germans, who lived in the Sudetenland region bordering Germany and Austria, to demand self-rule. He threatened to declare war on Czechoslovakia if this demand were not met. In 1938, under the terms of the Munich Agreement, Czechoslovakia was forced to give up the Sudetenland to Germany.

| | |
|---|---|
| c. A.D. 500 | Ancestors of the Czechs settle in Bohemia and Moravia. |
| 800's | Slavic tribes unite to form the Great Moravian Empire. |
| c. A.D. 900 | Bohemia develops into a powerful central European nation ruled by Přemyslid family. |
| 1347 | Charles IV of Bohemia becomes Holy Roman Emperor. |
| 1348 | Prague University is founded. |
| 1415 | Religious reformer John Hus is found guilty of heresy and is burned at the stake. |
| 1419–1436 | The Hussite Wars bring turmoil to Bohemia. |
| 1526 | The Habsburgs take control of Bohemia. |
| 1618 | A Czech revolt touches off the Thirty Years' War. |
| 1620 | The Habsburg Army defeats the Czechs in the Battle of White Mountain, and Bohemia loses its self-governing powers. |
| Mid-1800's | Czech nationalistic feelings grow, and a movement for self-government takes shape. |
| 1918 | The Czechs and Slovaks form Czechoslovakia as an independent nation. |
| 1938 | Munich Agreement turns over the Sudetenland to Germany. |
| 1939 | German armies occupy Czechoslovakia. |
| 1945 | Soviet troops free most of Czechoslovakia, while U.S. troops liberate parts of Bohemia. |
| 1948 | Communists take over the government. |
| 1968 | The Prague Spring, a period of liberal reform, ends with the invasion of Czechoslovakia by troops from the Soviet Union and its allies. |
| 1969 | Gustáv Husák becomes Communist Party leader in Czechoslovakia after Alexander Dubček is removed from office. |
| 1989 | Communist regime collapses, and Husák is replaced as president of Czechoslovakia by Václav Havel. |
| 1991 | Soviet Union withdraws its troops from Czechoslovakia. |
| 1992 | Havel resigns as president as Czech and Slovak leaders prepare to dissolve the Czechoslav federation. |
| 1993 | The Czech Republic is formally established on Jan. 1; Havel returns to power as president of the new republic. |
| 1999 | The Czech Republic joins the North Atlantic Treaty Organization. |
| 2004 | The Czech Republic joins the European Union. |
| 2011 | Vaclav Havel dies. |

John Hus
(1369?-1415)

Bedřich Smetana
(1824-1884)

Václav Havel
(1936-2011)

A young Czech hammers at the armor of a Soviet tank during the invasion of Czechoslovakia in 1968. The Soviet Union and its Eastern European allies sent troops to crush the Czechoslovak movement toward democracy known as the Prague Spring.

In 1939, German troops invaded Czechoslovakia, and Bohemia and Moravia became a German protectorate, while Slovakia became a separate republic under German control. By 1945, Allied troops had freed Czechoslovakia from the Germans. During the years that followed, Czechoslovakia fell under Communist rule. In 1948, the Communists forced the president of Czechoslovakia to form a new government consisting entirely of Communists.

In 1968, the Communist Party appointed Alexander Dubček as party leader. Under Dubček, the government introduced a program of liberal reforms, known as the *Prague Spring*. However, the Soviet Union and other Eastern European nations, fearing that the reforms would weaken Communist control, invaded the country in August 1968. Gustáv Husák relaced Dubček as party leader in 1969 and became president of the country in 1975. Under Husák's leadership, Czechoslovakia remained a Communist state until 1989.

Surrounded by his fellow dignitaries, Václav Havel takes the oath of office in an inauguration ceremony held Feb. 2, 1993. Havel was elected the first president of the newly independent Czech Republic after stepping down as president of Czechoslovakia shortly before the federation was dissolved.

# PRAGUE

Prague, situated on both banks of the Vltava River, is the capital and largest city of the Czech Republic. It is also an important center of culture and learning. One of central Europe's oldest and most beautiful cities, Prague is often called the *Golden City*. Its picturesque stone castles set amid rolling hills could be the setting for a fairy tale.

Because Prague escaped massive bombing during World War II (1939–1945), the beauty and historic significance of many of its old buildings and churches are virtually unrivaled throughout Europe. Many historic events took place in Prague, including the uprising that triggered the Thirty Years' War in 1618 and the Soviet invasion of 1968, when armored tanks ended the brief reform movement known as the Prague Spring.

## Early history

According to legend, Prague was founded about A.D. 800 by Princess Libuše, who had a vision of a glorious city while standing on a rocky ledge above the east bank of the River Vltava. The royal Palace of Vyšehrad was built on that spot in the 800's. However, the original settlement of Prague took place on the opposite bank of the river. At the end of the 900's, the first Czech kings built a massive hilltop citadel on *Hradčany* (Castle Hill), which dominates the city's skyline.

The commercial district of *Staré Město* (Old Town) developed across the river from Castle Hill, at the intersection of important trade routes. German colonists settled in this section starting in the early 1200's. A large and flourishing Jewish community in this district was confined—beginning in the 1200's—inside a walled ghetto. The walls were pulled down in the mid-1800's, but one of the ghetto's original synagogues has survived.

In 1257, south of Hradčany and within the outer wall of the citadel, King Otakar II

Waltzing couples in a Prague park reflect the graceful, elegant atmosphere of the Czech Republic's historic capital. The city was probably founded in the A.D. 800's and later served as the residence of Bohemian kings.

Houses along the streets of Malá Strana recall Bohemia's golden age. Many other buildings in Prague that were constructed under the post-World War II Communist government are now in poor condition.

founded the Lesser Town of Prague—later known as the *Malá Strana,* or Lesser Quarter. Many beautiful old palaces line the narrow, winding streets of Malá Strana.

In 1348, Charles IV established the *Nové Město* (New Town), which extends south of the Staré Město and down to Vyšehrad. Today, the Nové Město is Prague's business district, featuring Wenceslas Square—actually a wide boulevard lined with hotels and shops.

## The golden age

Prague soon grew into an important trading center, and in time it became the residence of the Bohemian kings. Charles IV, who also ruled the Holy Roman Empire, built many impressive buildings in the city.

Under Charles IV, Prague emerged as a cultural center of international renown. In 1348, the king founded Charles University, the oldest university in central Europe, and his court included the Italian poet Petrarch.

The French architect Matthias of Arras and German
architect Peter Parler transformed St. Vitus' Cathedral into an architectural wonder.

During the 1600's, religious wars and conflict
between the Habsburg monarchs and the Czech nobility caused great damage to the city. The triumph
of the Habsburgs in the Thirty Years' War was followed by an ambitious rebuilding campaign that
gave the city its present appearance. Unlike the
neighboring cities of Vienna, Austria, and Bu-
dapest, Hungary—which boast wide boulevards
and huge apartment blocks—the old center of
Prague remains unchanged from days gone by.

Prague's beauty, culture, and history make it popular with tourists. The tourism industry has expanded greatly in Prague and the rest of the Czech
Republic since the late 1980's.

High above the Charles Bridge, the spires of St.
Vitus' Cathedral crown Prague's Hradčany (Castle
Hill). The Hradčany is also the site of Prague Castle,
which now serves as an art museum and the official
residence of the Czech Republic's president.

# ART AND POLITICS

The line between art and life is a thin one in the Czech Republic—a country whose first elected president was a noted playwright. Since before World War I (1914–1918), Czech writers have used wit and imagination to reveal many truths about politics, authority, and the human condition. During the years of Communist rule, when freedom of expression was greatly restricted, their writings challenged the system, poking fun at its often absurd ways.

## Hašek's immortal Schwejk

In the days before World War I, the writings of Jaroslav Hašek subtly attacked Austrian rule of the Czech people through a shabbily dressed character named Schwejk. In Hašek's tales, Schwejk is thrown out of the army as an imbecile, but is drafted after the outbreak of World War I. Although he appears stupid, Schwejk succeeds in ridiculing and disrupting the whole military establishment.

To Hašek, Schwejk was a man whose humble achievements were ultimately more important than those of the greatest military figures, and the character became a hero of modern literature. Hašek himself was a colorful personality who enjoyed practical jokes. On one occasion, he identified himself as a "spy" in the guest register of a leading Prague hotel. Within moments, soldiers surrounded the hotel, only to find that the "spy" was none other than Hašek. When asked why he did it, Hašek replied that he was simply testing the efficiency of the Austrian secret service.

After World War I (1914–1918), the scholar and statesman Tomáš Masaryk became the first president of the newly independent state of Czechoslovakia. Masaryk's government gave enormous support to the arts, but the Germans squelched all freedom of expression during their occupation of

## THE CZECH TRADITION

**Tomáš Masaryk became the first president of the newly independent state of Czechoslovakia in 1918. Masaryk was a scholar and philosopher who did not enter politics until he was almost 40 years old. His academic background and the encouragement he gave to the arts reflect** the traditional Czech connection between politics and the arts. Chief among the writers who used humor to ridicule government failings was Jaroslav Hašek. His beloved character, Schwejk, who outwitted the hard-hearted military authorities of World War I, was a model for later comedic characters such as Sergeant Bilko, a hero of American television comedy. When the Communists took over the Czech state in 1948, this tradition of ridiculing authority continued in the works of such writers as Bohumil Hrabal, whose novel *I Served the King of England* was published underground in the late 1950's. Jazz, officially dismissed by the Nazis and the Communists as a symbol of moral decay, became a Czech symbol of liberty and inspired Josef Skvorecky's highly influential 1960's novella *The Bass Saxophone*.

Czech films combine humor and social comment in Jiří Menzel's *Closely Watched Trains* (1966) and Milos Forman's *The Firemen's Ball* (1967). Both films satirize authority and small-town life.

During the Prague Spring of 1968, those who disagreed with Communist policies protested against the government in massive demonstrations that were later crushed by Soviet troops.

In November 1989, crowds of students took to the streets of Prague and other Czechoslovakian cities, beginning the movement that came to be known as the "velvet revolution."

While it involved some confrontation with government forces, the movement was largely peaceful. Artists, writers, and even rock musicians played a key role in the mass uprising. The velvet revolution brought down the Communist system in Czechoslovakia, paving the way not only for a democratic government system and free market economy, but also for a new freedom of expression for Czech writers and artists.

Many talented people were forced to leave Czechoslovakia after the Soviet invasion of 1968. However, the arts continued to play a major role in political developments. Charter 77—a manifesto of individual liberties signed by nearly all the country's remaining intellectuals and cultural figures—was developed in 1977 in protest of the imprisonment of an antiauthoritarian rock band called The Plastic People of the Universe.

The Magic Lantern Theater in Prague literally provided the stage for Czechoslovakia's "velvet revolution" in 1989. Founded in 1958, the theater was famous for its innovative plays and films, and it became a meeting place for Czechoslovakia's artists and writers. In December 1989, Václav Havel and Alexander Dubček emerged in triumph from its backstage rooms to proclaim the return of democracy to Czechoslovakia. Havel served as president of Czechoslovakia from 1989 to 1992. He stepped down briefly during the breakup of the country. He then served two terms as president of the Czech Republic, from 1993 to 2003.

the country in World War II. After the war, the Communist rulers put equally tight restrictions on the arts, severely punishing anyone who dared to oppose them.

## The coming of Prague Spring

The bureaucratic absurdities of Czechoslovakia's Communist leaders provided rich subject matter for Czech writers, whose sense of humor helped them cope with the rigid authorities. However, many fine writers were rarely published during those years because their work did not conform to official policy.

Novelists, playwrights, musicians, and filmmakers joined forces in the 1960's to create a mood of optimism and vitality. Prague's Theater at the Balustrade, where Václav Havel's plays were first performed, became a rallying point for Czechs who disagreed with Communist policies. The new spirit culminated in the Prague Spring of 1968.

Gifted filmmakers brought the great Czech spirit of expression to worldwide attention. Jiri Menzel's film *Closely Watched Trains* won an Academy Award for Best Foreign Film of 1966. The film, based on a novella by Bohumil Hrabal, told the story of a young railway employee coming of age during the last days of World War II.

Václav Havel (right) celebrates the triumph of democracy with Alexander Dubček, whose policies led to the Prague Spring.

# ENVIRONMENT

Although the Czech Republic is completely landlocked, two important rivers link the country to the Baltic and North seas. The Elbe River, one of the major commercial waterways of central Europe, rises in the northern part of the republic and flows northwest through Germany before emptying into the North Sea. The Oder River rises in the Sudeten Mountains and flows northward to the Baltic Sea.

The Czech Republic can be divided into five major land regions. These are the Bohemian Mountains, the Sudeten Mountains, the Bohemian Basin, the Bohemian-Moravian Highlands, and the Moravian Lowlands.

## Mountains and forests

The Bohemian Mountain region, in the extreme west of the country, includes the Ore Mountains in the northwest and the Bohemian Forest in the southwest. Many people come to bathe in the mineral springs of the Ore Mountains. The mountains—rising more than 2,500 feet (762 meters) above sea level—make the region a popular skiing destination.

The Bohemian Mountain region contains large deposits of coal and uranium ore, making it a major center for the Czech Republic's mining and manufacturing industries. Tourists enjoy wandering along the region's forest paths—a route taken by highway robbers long ago. The thickly wooded hills of the Bo-

The picturesque town of Mikulov lies on the edge of the Moravian Lowlands. One of the Czech Republic's leading industrial and mining centers lies to the northeast, around the city of Ostrava.

The rugged cliffs of these mountains in Bohemia rise in dramatic contrast to the rolling hills beyond them. The Bohemian and Sudeten mountains cover much of the northern Czech Republic.

hemian Forest are an important source of lumber and wood products.

The higher and more rugged Sudeten Mountains in the north-central part of the Czech Republic define the country's border with Poland. This region includes industrial cities and towns as well as farming areas, forests, and resorts.

South of the Sudeten Mountains lies the Bohemian Basin, a low-lying area of plains and rolling hills. In this fertile region, farmers grow potatoes, rye, sugar beets, wheat, and other crops. Major rivers include the Vltava, which flows northward toward Prague, and the Elbe, which flows westward across the basin.

## Farms and factories

The Bohemian Basin is also home to some of the Czech Republic's major industrial cities. Prague and Hradec Králové are among the region's industrial centers.

The Bohemian-Moravian Highlands cover most of southern Bohemia—a region of high plains, low hills, and plateaus. Numerous small towns and villages are

A street vendor in Prague sells fresh fruits and vegetables grown in the surrounding countryside. The city lies in the Bohemian Basin, where some of the country's most fertile farmland can be found.

Plzen, situated in the western part of the republic, ranks as one of the nation's major industrial centers. Its factories manufacture automobiles, buses, machinery, military equipment, and trucks, while breweries produce world-famous beer.

A craft worker in Karlovy Vary (Karlsbad) inscribes a design onto a glass bowl. Northern Bohemia has been a leading center of the glass industry since the 1600's.

scattered throughout this area, which consists mainly of farmland.

The densely populated Moravian Lowlands extend across the southeastern part of the Czech Republic. Fertile farmland covers the Morava River Valley, where farmers raise livestock and grow such crops as corn, potatoes, rye, sugar beets, and wheat.

The Moravian Lowlands also have the nation's most important coal fields. One of the Czech Republic's leading industrial and mining centers is situated around the city of Ostrava. Plzen, the largest city in the area, is a manufacturing center known for automaking and beer brewing.

Brno, an important industrial city near the country's southeastern border with Slovakia, is the Czech Republic's second largest city and the chief city of Moravia. Founded in the 800's, Brno is now a manufacturing center that produces automobiles, chemicals, iron and steel, leather goods, machinery, and textiles.

# DENMARK

Denmark is a small kingdom in northern Europe that is almost completely surrounded by water. The Jutland Peninsula accounts for almost 70 percent of the land, and Denmark also has 482 nearby islands. More than half of the Danes live on the islands near the peninsula.

Although it lacks natural resources and ranks as one of the smallest countries in Europe both in area and population, Denmark is a thriving, prosperous nation. Its people enjoy one of the highest standards of living in the world. More than 85 percent of the Danes live in busy, modern cities. However, visitors are drawn to the charming Danish countryside, where castles and windmills rise above the rolling landscape and picturesque houses stand amid well-kept farmlands.

Although Denmark is still known as a great shipping and fishing nation, service industries and manufacturing are now its leading industries. Danish factories produce high-quality goods including furniture, porcelain, silverware, stereos, and television sets. The sea around the country's islands provides a cheap way for Denmark to import its industrial needs and export its products.

The Danish people share many cultural traits with their neighbors in Sweden and Norway. The Danish language is quite similar to Swedish and Norwegian tongues, and, like their Scandinavian neighbors, most Danes belong to the Evangelical Lutheran Church, the official church of Denmark.

The nation's highly developed welfare program also resembles the systems of Sweden and Norway. Denmark was one of the first countries in the world to establish widespread social services by introducing public relief for the sick, unemployed, and aged.

The bond between the Scandinavian countries reached a peak in 1397, when Denmark's Queen Margaret united Denmark, Norway, and Sweden in the Union of Kalmar. However, many battles have taken place between these nations. During the 1600's and 1700's, Sweden defeated Denmark in several wars over control of the Baltic Sea.

## FACTS

| | |
|---|---|
| **Official name:** | **Kongeriget Danmark (Kingdom of Denmark)** |
| **Capital:** | **Copenhagen** |
| **Terrain:** | **Low and flat to gently rolling plains** |
| **Area:** | **16,639 mi² (43,094 km²)** |
| **Climate:** | **Temperate; humid and overcast; mild, windy winters and cool summers** |
| **Main rivers:** | **Guden, Skjern** |
| **Highest elevation:** | **Yding Skovhøj, 568 ft (173 m)** |
| **Lowest elevation:** | **Sea level along the coasts** |
| **Form of government:** | **Constitutional monarchy** |
| **Head of state:** | **Monarch** |
| **Head of government:** | **Prime minister** |
| **Administrative areas:** | **5 regions; Greenland is a province of Denmark; the Faroe Islands are a self-governing administrative division** |
| **Legislature:** | **Folketing (Parliament) with 179 members serving four-year terms** |
| **Court system:** | **Supreme Court** |
| **Armed forces:** | **29,600 troops** |
| **National holiday:** | **Constitution Day - June 5 (1849)** |
| **Estimated 2010 population:** | **5,476,000** |
| **Population density:** | **329 persons per mi² (127 per km²)** |
| **Population distribution:** | **86% urban, 14% rural** |
| **Life expectancy in years:** | **Male, 76; female, 80** |
| **Doctors per 1,000 people:** | **3.6** |
| **Birth rate per 1,000:** | **12** |
| **Death rate per 1,000:** | **10** |
| **Infant mortality:** | **4 deaths per 1,000 live births** |
| **Age structure:** | **0-14: 18%; 15-64: 66%; 65 and over: 16%** |
| **Internet users per 100 people:** | **85** |
| **Internet code:** | **.dk** |
| **Languages spoken:** | **Danish, English, Faroese, Greenlandic, German** |
| **Religions:** | **Evangelical Lutheran 95%, other Christian 3%, Muslim 2%** |
| **Currency:** | **Danish krone** |
| **Gross domestic product (GDP) in 2008:** | **$342.93 billion U.S.** |
| **Real annual growth rate (2008):** | **-0.6%** |
| **GDP per capita (2008):** | **$62,772 U.S.** |
| **Goods exported:** | **Machinery, meat and other food products, petroleum products, pharmaceuticals** |
| **Goods imported:** | **Consumer goods, food, machinery, transportation equipment** |
| **Trading partners:** | **Germany, Netherlands, Norway, Sweden, United Kingdom, United States** |

Denmark consists of the Jutland Peninsula and hundreds of islands to the east. The peninsula projects north from mainland Europe, toward the other Scandinavian nations. The islands lie between the Kattegat, a narrow extension of the North Sea, and the Baltic Sea.

A soldier of the Royal Guard parades outside the Amalienborg Palace in Copenhagen.

In centuries past, Denmark held considerable power over neighboring lands. The nation's long tradition of expansion began in the early 800's, when Danish Vikings raided and burned towns on the coasts of what are now Belgium, France, and the Netherlands and sailed away with slaves and treasure. In 865, the Danes invaded England and settled in the eastern half of the country.

Denmark chose to remain neutral in World War I (1914–1918). Early in World War II (1939–1945), German troops invaded Denmark, and the Danes surrendered after a few hours of fighting. After the war, U.S. aid helped the Danes rebuild industries that had been damaged.

Today, Denmark's economy remains strong by worldwide standards. However, the Danish economy is greatly affected by international trends and developments, since Denmark must sell its products to other countries to pay for the fuels and metals it imports. Denmark belongs to the European Union, but it reserves the right to abstain from a common currency or defense policy. Denmark is a constitutional monarchy, with a prime minister as the head of government.

# ENVIRONMENT

The landscape of Denmark owes most of its features to the Ice Age, when glaciers moved slowly over the region from about 2 million to 10,000 years ago. The advancing glaciers moved rocks and boulders with them, and many were crushed almost to powder. This debris formed *moraines* (banks and ridges) where the ice stopped.

These glacial deposits almost completely cover the flat layers of limestone that form the bedrock of the Jutland Peninsula and the islands. White, finely grained limestone formations are visible in only a few places, such as Lim Fiord and the islands of Møn and Bornholm.

## Land regions

The smooth curves of Denmark's Western Dune Coast consist chiefly of great sandy beaches that close off many fiords once connected to the sea. Inland, the Western Sand Plains were formed when melting glaciers flowed over the land, depositing quantities of sand.

The East-Central Hills, which include much of Jutland and almost all the nearby islands, make up Denmark's largest land region, and the deep moraine soils on these islands provide the best farmland in Denmark. Here, Danish farmers grow such crops as barley, potatoes, sugar beets, and *rape* (a leafy herb).

Esbjerg, with its huge harbor on the North Sea coast, is Denmark's major fishing port and a major gateway for imports and exports. The city also serves as a center of operations for Denmark's North Sea oil exploration.

Most of the crops grown on these islands are used for animal feed, since livestock production is the major activity on most Danish farms. More than two-thirds of the country's agricultural products are exported, and Denmark is world famous for its meat and dairy products, including cheeses, butter, bacon, and ham.

The Lim Fiord winds through northern Jutland for 112 miles (180 kilometers), forming an inland lagoon 15 miles (24 kilometers) wide. A beach on the Western Dune Coast closes the fiord's outlet to the North Sea, so small vessels use the Thyborøn Canal to travel between the fiord and the sea.

Sjælland, the nation's largest island and the most thickly populated part of Denmark, lies in the East-Central Hills. Most of Copenhagen, Denmark's capital and largest city, stands on this island. Many

Dairy cows graze on one of Denmark's many farms. More than half of the country's land is devoted to agriculture. Producing dairy products is the major activity on many Danish farms.

The sheer cliffs carved by the sea from cream-colored chalk (finely grained limestone) are a famous landmark on the island of Møn. Known as Møns Klint, the cliffs tower 400 feet (120 meters) above a narrow beach that can be reached only by steep steps.

Fishing boats are pulled up for the night onto the smooth, dune-lined beaches of western Jutland.

Except for the extreme southeast of the island of Bornholm, Denmark consists of a glacial deposit over a limestone base. The landscape is made up of small hills, moors, ridges, hilly islands, and raised seabed.

of the nation's industries are located in the Copenhagen area.

Located on the northeast coast of Jutland, the Northern Flat Plains were once part of the seabed. The region rose up from the water when the weight of the ancient glaciers was lifted as they melted. Like the East-Central Hills, the Northern Flat Plains have deep, rich moraine soil and many farms.

## Natural resources

For hundreds of years, windmills have been a prominent feature in the Danish landscape. The mills make use of the wind to turn grindstones and other agricultural devices. Today, windmills generate about 20 percent of Denmark's electric power.

Denmark has few other natural resources. Although the country gets some natural gas and petroleum from wells in the North Sea, it must still import petroleum products. Other mined products include chalk and industrial clay. Because the land is flat or gently rolling, the rivers cannot be used to generate hydroelectric power. However, the sea, which almost surrounds Denmark, is rich in such fish as cod, herring, Norway pout, sand lances, sprat, and whiting.

## Climate

Mainly because it is almost surrounded by water, Denmark has a mild, damp climate that is affected by sea winds throughout the year. In winter, the sea is warmer than the land, and in summer, it is cooler. As a result, west winds from the sea warm Denmark in winter and cool the land in summer.

The climate varies little throughout the country because Denmark has no mountains to block the sea winds. However, the moisture-bearing sea winds reach western Denmark first, so that area receives more rainfall than eastern Denmark. Snow falls from 20 to 30 days a year, but it usually melts quickly. Fog and mist occur frequently, especially on the west coast in winter.

Denmark's coastal waters are rich fishing grounds that provide a bountiful catch.

# COPENHAGEN

Copenhagen, Denmark's capital and largest city, traces its origins to the mid-1000's, when it was a small fishing village. Today, the city is Denmark's major port and also serves as the center of the nation's economic, political, and cultural activity. Most of the city lies on the east coast of the island of Sjælland, while other sections are on Amager, an island just east of Sjælland.

## A historic city

Until the mid-1100's, Copenhagen—whose well-sheltered port provides immediate access to the sea—was continually raided by pirates. In 1167, Archbishop Absalon of Roskilde built a castle to protect the harbor, which encouraged the growth and development of Copenhagen as a trade center. In 1254, Copenhagen was granted a town charter.

As the town continued to grow in size and importance within the castle's protective walls, it attracted the jealous wrath of German Hanseatic

Town Hall Square, at the center of Copenhagen, is surrounded by open-air cafes and historic buildings, including the Town Hall, which was built in the late 1800's. The Dragon Fountain, in the center of the square, is dedicated to Hans Christian Andersen, Denmark's most famous writer. Andersen's fairy tales have enchanted generations of the world's children.

merchants, who destroyed the castle in 1369. The settlement was rebuilt, and in 1443, it became the capital of Denmark.

During the next 400 years, Copenhagen suffered through terrible times. In the 1600's, the town was attacked by Swedish forces. In the 1700's, much of it was destroyed by fires, while epidemics killed many of its people. The town had barely recovered before being bombarded by the British fleet in 1801 and 1807, during the Napoleonic Wars.

Copenhagen bounced back each time from the devastation, and it continued to grow as an economic, military, and political center. During the 1850's, the city expanded to the north and west, and in the late 1800's, it also experienced rapid economic growth and began to develop industries.

Picturesque houses built by wealthy Danish merchants during the 1700's and 1800's line the Nyhavn Canal. The Nyhavn area features numerous sailing ships, historic buildings, and fashionable cafes.

*The Little Mermaid,* a bronze statue of one of Hans Christian Andersen's most beloved characters, watches over Copenhagen's harbor.

1843. Millions of visitors come to Tivoli every summer to enjoy its various forms of entertainment, including shows, games, rides, restaurants, museums, and vast flower gardens. Tivoli also boasts a concert hall and pantomime theater. On certain nights, the park closes with a magnificent fireworks display.

East of the Tivoli Gardens stands Christiansborg Castle, the seat of the Danish Parliament (Folketing) and Supreme Court. The castle stands on the same site as the castle built by Archbishop Absalon, and the ruins of the original castle can be seen under the present building. Farther east stands Amalienborg Palace, a square of four mansions that has served as the royal residence since 1794.

In addition to its historic buildings, Copenhagen boasts a wealth of fascinating museums. The National Museum, built in the 1740's, houses one of the world's finest collections of Stone Age tools, as well as runic stones dating from Viking times. The Museum of Decorative Arts displays a collection of ceramics, silverware, and Flemish tapestries from the Middle Ages.

## Modern Copenhagen

Considered by many to be one of the most delightful capitals in the world, present-day Copenhagen boasts a wealth of beautiful, historic buildings, many of them built in the 1600's, during the reign of King Christian IV.

Christian IV became known as Denmark's greatest builder. He sponsored the construction of some of Copenhagen's most elegant buildings and even helped design some of them, including the Rosenborg Palace (1606) and the Stock Exchange (1624).

The Rosenborg Palace, once the king's spring and autumn residence, is now a historical museum containing a breathtaking display of all the Danish crown jewels, as well as the personal effects of Danish kings from the time of Christian IV. The Stock Exchange, the oldest building of its kind still used for its original purpose, is remarkable for its spire adorned with intertwined dragons' tails.

Another distinctive feature of Copenhagen is the world-famous Tivoli Gardens amusement park, built in

The Tivoli Gardens amusement park, in the center of the city, is a popular meeting place for Danes of all ages. The Chinese pagoda, lit by hundreds of tiny lights, is one of the park's many popular attractions.

# DJIBOUTI

Djibouti is a small country on the eastern coast of Africa, where the continent nearly touches the Arabian Peninsula. Its hot, dry climate, near-barren land, and lack of resources have left it poor and underdeveloped. The country has very little industry. The only agricultural activity is livestock herding.

Djibouti's main advantage is its location, which has helped make its capital city—also named Djibouti—a major port. A railroad linking Djibouti with Ethiopia also helps make the capital a center for trade. The economy of the entire nation depends almost totally on this sea and rail trade.

In addition, Djibouti's location is important in world politics. If a powerful nation ever gained control of this area, it could interfere with the many ships that now move freely past Djibouti's coast traveling through the Suez Canal and the Red Sea between the Mediterranean Sea and the Indian Ocean.

Despite Djibouti's desert climate and desolate land, people have lived in the area since prehistoric times. During the A.D. 800's, Muslims from Arabia converted the Afars, a nomadic group living in the region, to Islam. The Afars established several Islamic states and fought wars with the Christians in neighboring Ethiopia from the 1200's through the early 1600's.

By the 1800's, however, a group of Somali nomads from the south, called the Issas, had taken over a large part of the Afars' grazing lands. The two groups grew hostile toward each other.

In 1862, France bought a port in the region and later established a coaling station for its ships. The French eventually gained control of more land and turned it into a territory named French Somaliland. In 1967, the name was changed to the French Territory of the Afars and Issas.

Opposition to French rule grew in the 1970's, especially from the Issas. In May 1977, the people voted overwhelmingly for independence. On June 27, the territory became the independent nation of Djibouti.

## FACTS

| | |
|---|---|
| Official name: | Republic of Djibouti |
| Capital: | Djibouti |
| Terrain: | Coastal plain and plateau separated by central mountains |
| Area: | 8,958 mi$^2$ (23,200 km$^2$) |
| Climate: | Desert; torrid, dry |
| Main rivers: | N/A |
| Highest elevation: | Mousaalli, 6,768 ft (2,063 m) |
| Lowest elevation: | Lake Assal, 509 ft (155 m) below sea level |
| Form of government: | Republic |
| Head of state: | President |
| Head of government: | Prime minister |
| Administrative areas: | 5 cercles (districts) |
| Legislature: | Chambre des Deputes (Chamber of Deputies) with 65 members serving five-year terms |
| Court system: | Cour Supreme (Supreme Court) |
| Armed forces: | 10,500 troops |
| National holiday: | Independence Day - June 27 (1977) |
| Estimated 2010 population: | 877,000 |
| Population density: | 98 persons per mi$^2$ (38 per km$^2$) |
| Population distribution: | 88% urban, 12% rural |
| Life expectancy in years: | Male, 46; female, 50 |
| Doctors per 1,000 people: | 0.2 |
| Birth rate per 1,000: | 34 |
| Death rate per 1,000: | 16 |
| Infant mortality: | 83 deaths per 1,000 live births |
| Age structure: | 0-14: 41%; 15-64: 56%; 65 and over: 3% |
| Internet users per 100 people: | 1.4 |
| Internet code: | .dj |
| Languages spoken: | French (official), Arabic (official), Somali, Afar |
| Religions: | Muslim 94%, Christian 6% |
| Currency: | Djiboutian franc |
| Gross domestic product (GDP) in 2008: | $973 million U.S. |
| Real annual growth rate (2008): | 6.0% |
| GDP per capita (2008): | $1,161 U.S. |
| Goods exported: | Coffee, hides and skins, re-exports |
| Goods imported: | Food and beverages, machinery, petroleum products, transportation equipment |
| Trading partners: | Ethiopia, Saudi Arabia, Somalia |

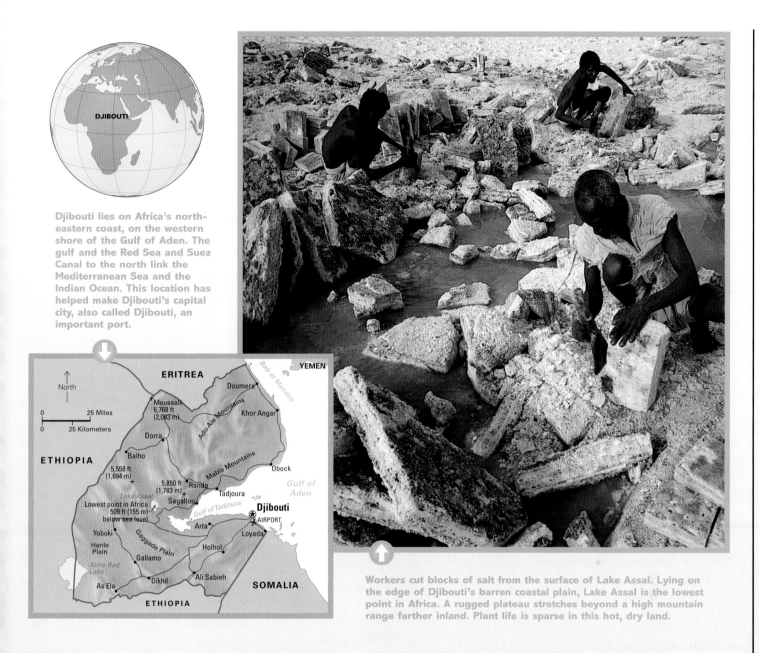

Djibouti lies on Africa's northeastern coast, on the western shore of the Gulf of Aden. The gulf and the Red Sea and Suez Canal to the north link the Mediterranean Sea and the Indian Ocean. This location has helped make Djibouti's capital city, also called Djibouti, an important port.

Workers cut blocks of salt from the surface of Lake Assal. Lying on the edge of Djibouti's barren coastal plain, Lake Assal is the lowest point in Africa. A rugged plateau stretches beyond a high mountain range farther inland. Plant life is sparse in this hot, dry land.

The Afars and the Issas still make up the two main ethnic groups in Djibouti. The Afars live in the north and west, and the Issas live in the south. In 1991, government forces dominated by Issas began fighting rebel Afar guerrillas. In 1994, the two sides signed a peace agreement. However, some Afar rebels continued fighting against the government. In 2000, the last remaining rebels and the government signed a peace agreement.

Many Afars and Issas speak Afar or Somali rather than Djibouti's official languages of French and Arabic. Many also follow the nomadic way of life of their ancestors. They wander over the desolate land with their herds of animals. Scorching heat, little water, and scarce grazing land make life difficult for the nomads, so large numbers of Afars and Issas have moved to the capital city of Djibouti in search of a better life. However, poverty plagues these urban dwellers, and high unemployment remains a major problem.

Educational opportunities are limited, but about two-thirds of the adult population can read and write. Large numbers of Djibouti workers spend much of their income on *khat,* a leaf that produces a feeling of well-being when it is chewed.

# DOMINICA

A small island republic in the Lesser Antilles, Dominica lies 320 miles (515 kilometers) north of Venezuela. This mountainous, tree-covered island, which was formed by volcanic eruptions, is one of the most unspoiled islands in the Caribbean.

## An unspoiled wilderness

Traveling around Dominica can be difficult because some roads are so poor and others have been damaged by heavy rainfall and avalanches. But visitors willing to brave the bumpy, desolate roads are rewarded with some of the most magnificent, unspoiled scenery in the West Indies. A ride across the island is an extraordinary journey through dense rain forests and over high, volcanic mountains.

The rain forests of the interior support a large variety of trees, including cedar and mahogany. In addition, the forests contain abundant bird life. Flowering plants, such as hibiscus, lilies, and orchids, also blanket the land.

Rain water tumbles down the mountains in huge torrents and merges with swift-flowing rivers. The rivers, most of which are too rough to be used by boats, flow into coastal bays lined by black volcanic sands.

Dominica's interior also features hundreds of waterfalls and crystal-clear lakes. Boiling Lake, with its bubbling sulfur springs, is the second largest such lake in the world.

## History and government

Arawak Indians, the first people to settle in Dominica, arrived there about 2,000 years ago. About 1,000 years later, Carib Indians took over the island. Christopher Columbus sighted the island on Sunday, Nov. 3, 1493. He named the island *Dominica,* which is the Latin word for *Sunday.*

French and British settlers began to arrive in Dominica in the 1600's, and the French, British, and Carib fought for control of the island for

## FACTS

| | |
|---|---|
| Official name: | Commonwealth of Dominica |
| Capital: | Roseau |
| Terrain: | Rugged mountains of volcanic origin |
| Area: | 290 mi² (751 km²) |
| Climate: | Tropical; moderated by northeast trade winds; heavy rainfall |
| Main river: | Layou |
| Highest elevation: | Morne Diablotin, 4,747 ft (1,447 m) |
| Lowest elevation: | Caribbean Sea, sea level |
| Form of government: | Republic |
| Head of state: | President |
| Head of government: | Prime minister |
| Administrative areas: | 10 parishes |
| Legislature: | House of Assembly with 30 members serving five-year terms |
| Court system: | Eastern Caribbean Supreme Court |
| Armed forces: | N/A |
| National holiday: | Independence Day - November 3 (1978) |
| Estimated 2010 population: | 73,000 |
| Population density: | 252 persons per mi² (97 per km²) |
| Population distribution: | 74% urban, 26% rural |
| Life expectancy in years: | Male, 72; female, 78 |
| Doctors per 1,000 people: | 0.5 |
| Birth rate per 1,000: | 16 |
| Death rate per 1,000: | 8 |
| Infant mortality: | 15 deaths per 1,000 live births |
| Age structure: | 0-14: 27%; 15-64: 63%; 65 and over: 10% |
| Internet users per 100 people: | 39 |
| Internet code: | .dm |
| Languages spoken: | English (official), French patois |
| Religions: | Roman Catholic 61.4%, other Christian 29.5%, Rastafarian 1.3%, other 7.8% |
| Currency: | East Caribbean dollar |
| Gross domestic product (GDP) in 2008: | $364 million U.S. |
| Real annual growth rate (2008): | 2.6% |
| GDP per capita (2008): | $4,554 U.S. |
| Goods exported: | Bananas, soap, vegetables |
| Goods imported: | Chemicals, food, machinery, motor vehicles, petroleum products |
| Trading partners: | China, Jamaica, Japan, Trinidad and Tobago, United States |

many years. It was not until 1763 that the British gained possession of Dominica. They established large plantations and brought in African slaves to work on them. The slaves were freed in 1834.

Between the 1930's and the 1970's, the United Kingdom gradually gave the island control over its internal affairs. Dominica became an independent republic on Nov. 3, 1978—exactly 485 years after Columbus first sighted the island from the deck of his ship.

A president is officially the country's chief executive, but a prime minister is the most power-ful official. The prime minister is a member of the Cabinet, which conducts the operations of the government. A legislature called the House of Assembly makes the nation's laws.

Today, Dominica is one of the poorest nations in the West Indies, and its many problems include poverty, economic mismanagement, and political corruption. A rugged landscape, heavy rainfall, avalanches, and hurricanes have added to the nation's difficulties. In spite of their country's problems, however, the people of Dominica welcome visitors with warmth and friendliness.

Dominica has been an independent republic since 1978. Most of its people have African or mixed African, British, and French ancestry. About three-fourths of the people live in urban areas, and the rest live in rural areas. English is spoken in the cities, while the villagers generally speak French patois—a mixture of African languages and French.

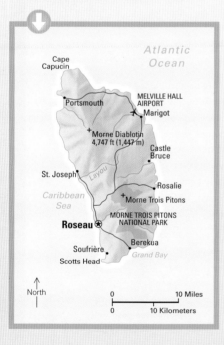

Volcanic mountains form the backdrop for a coastal scene on the island of Dominica. Some mountains in the north and south reach heights of more than 4,000 feet (1,200 meters).

# DOMINICAN REPUBLIC

The Dominican Republic is a mountainous country on the eastern two-thirds of the island of Hispaniola. It shares the island with Haiti to the west. The country lies in the West Indies island group, about 575 miles (925 kilometers) southeast of Miami, Florida. Santo Domingo, a busy port city of almost 2 million people, is its capital and largest city.

## The land

Mountains dominate the Dominican Republic. The country has the highest point in the West Indies—the 10,417-foot (3,175-meter) Duarte Peak. The *Cordillera Central,* or Central Mountain Range, runs northwest to southeast. The *Cordillera Septentrional,* or Northerly Range, is in the far north.

Between the two cordilleras lies the Cibao, an area of pine-covered slopes and a fertile plain called the Vega Real, or Royal Plain—the country's chief agricultural area.

In the dry west lie two other mountain ranges. Between them lies Lake Enriquillo, at 150 feet (46 meters) below sea level the lowest point in the West Indies. Most of the nation's sugar cane is grown in the less mountainous eastern region.

## The government

For much of its history, the Dominican Republic has been ruled by dictators or by other countries. Today, a president heads the country. The president is elected by the people for a four-year term and appoints a cabinet to help run the government and governors to head the provinces.

The people also elect the national legislature. It consists of a 32-member Senate and a 178-member Chamber of Deputies. Members of both houses are elected to four-year terms.

## FACTS

| | |
|---|---|
| Official name: | Republica Dominicana (Dominican Republic) |
| Capital: | Santo Domingo |
| Terrain: | Rugged highlands and mountains with fertile valleys interspersed |
| Area: | 18,815 mi² (48,730 km²) |
| Climate: | Tropical maritime; little seasonal temperature variation; seasonal variation in rainfall |
| Main rivers: | Yaque del Norte, Yaque del Sur, Yuna |
| Highest elevation: | Duarte Peak, 10,417 ft (3,175 m) |
| Lowest elevation: | Lake Enriquillo, 150 ft (46 m) below sea level |
| Form of government: | Democratic republic |
| Head of state: | President |
| Head of government: | President |
| Administrative areas: | 31 provincias (provinces), 1 distrito (district) |
| Legislature: | Congreso Nacional (National Congress) consisting of the Senado (Senate) with 32 members serving four-year terms and the Camara de Diputados (Chamber of Deputies) with 178 members serving four-year terms |
| Court system: | Corte Suprema (Supreme Court) |
| Armed forces: | 49,900 troops |
| National holiday: | Independence Day - February 27 (1844) |
| Estimated 2010 population: | 9,884,000 |
| Population density: | 525 persons per mi² (203 per km²) |
| Population distribution: | 68% urban, 32% rural |
| Life expectancy in years: | Male, 70; female, 75 |
| Doctors per 1,000 people: | 1.9 |
| Birth rate per 1,000: | 23 |
| Death rate per 1,000: | 6 |
| Infant mortality: | 31 deaths per 1,000 live births |
| Age structure: | 0-14: 33%; 15-64: 61%; 65 and over: 6% |
| Internet users per 100 people: | 26 |
| Internet code: | .do |
| Language spoken: | Spanish |
| Religions: | Roman Catholics 95%, other 5% |
| Currency: | Dominican peso |
| Gross domestic product (GDP) in 2008: | $45.69 billion U.S. |
| Real annual growth rate (2008): | 4.5% |
| GDP per capita (2008): | $4,918 U.S. |
| Goods exported: | Clothing, cocoa, ferronickel, sugar, tobacco |
| Goods imported: | Cotton, food, iron and steel, petroleum and petroleum product |
| Trading partners: | China, Japan, Mexico, Netherlands, United States |

# History

Christopher Columbus landed on the island of Hispaniola in 1492. Thousands of Spanish colonists soon followed, many seeking gold. In 1496, they founded La Nueva Isabela, now Santo Domingo, the first city in the Western Hemisphere founded by Europeans.

After the Spaniards nearly wiped out the Indians who were living on the island, many colonists left Hispaniola for more prosperous settlements. The king of Spain then ordered the remaining colonists to move to the Santo Domingo area. This action cost Spain its control of the northern and western sections of the island. France eventually took over the western end, now Haiti.

When black slaves in Haiti revolted, they gained control of the whole island. From 1822 to 1844, Haiti controlled all of Hispaniola. In 1844, Dominican heroes Juan Pablo Duarte, Francisco del Rosario Sánchez, and Ramón Mella led a successful revolt against the Haitians.

Spain governed the Dominican Republic from 1861 to 1865 to protect it from Haiti. Dictator Ulises (Lilis) Heureaux ruled the country from 1882 to 1899 and left it in debt. The United States took over the collection of customs duties from 1905 to 1941 and used the money to pay the debts. It also sent U.S. Marines to keep order in the Dominican Republic from 1916 to 1924.

Rafael Leonidas Trujillo Molina seized power in a military revolt in 1930. Trujillo ruled harshly for 30 years, allowing little freedom and imprisoning or killing his opponents. He ruled efficiently, and the country prospered economically. But the people did not benefit because profits went to the Trujillo family.

Conspirators shot and killed Trujillo in 1961. A power struggle then began among the military, the upper class, the people who wanted a democracy, and Communists. In 1965, the United States again sent troops to maintain order, to protect Americans there, and to keep Communists from taking over. Other countries sent troops too.

The last foreign troops left the Dominican Republic in 1966. Since that time, the country has had regular elections for its presidency.

In May 2004, flash floods from torrential rains caused widespread destruction in the Dominican Republic. About 680 people were killed or reported missing. Extensive deforestation contributed to the flooding.

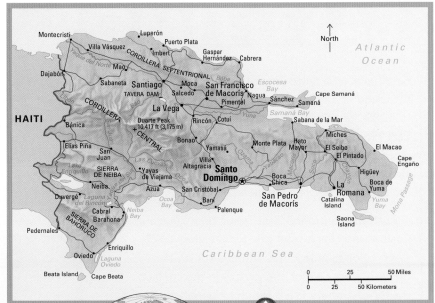

The Dominican Republic, a land of rich valleys and high, forested mountains, occupies the eastern two-thirds of the island of Hispaniola in the Caribbean Sea.

The rugged landscape of the Dominican Republic is dominated by the Cordillera Central, a high mountain range.

East Timor, also called Timor-Leste, is a country on the island of Timor in Southeast Asia. From 1975 to 1999, Indonesia claimed the region known as East Timor, but its authority there was never recognized by the United Nations (UN). In 1999, East Timor began transforming itself into an independent country. It became independent in 2002.

Portugal controlled East Timor, then known as Portuguese Timor, from the 1500's until 1975, when colonial authorities withdrew. A civil war then erupted. One of the parties in the conflict, the Revolutionary Front for an Independent East Timor (FRETILIN), declared East Timor's independence in November 1975. In December of that year, Indonesian forces invaded. In July 1976, Indonesia annexed East Timor as its 27th province.

Indonesia spent large sums of money in East Timor, but many of the people continued to resist Indonesian occupation. During the 1990's, the United States and other nations joined nongovernmental organizations in accusing Indonesia of serious human rights violations in East Timor.

## Peace efforts

José Ramos-Horta, an East Timorese politician and diplomat, shared the 1996 Nobel Prize for peace with Bishop Carlos Felipe Ximenes Belo for their efforts to secure a just settlement of the conflict.

Ramos-Horta became recognized as the main international spokesman for an independent East Timor. He became the special representative of the National Council of Maubere Resistance, later called the National Council of Timorese Resistance, a coalition of Timorese proindependence organizations. In 1998, he became vice president of the council. Two years later, he became the foreign minister of East Timor's transitional government.

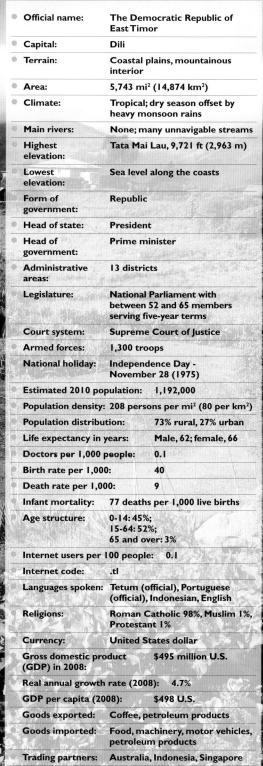

## FACTS

| | |
|---|---|
| Official name: | The Democratic Republic of East Timor |
| Capital: | Dili |
| Terrain: | Coastal plains, mountainous interior |
| Area: | 5,743 mi² (14,874 km²) |
| Climate: | Tropical; dry season offset by heavy monsoon rains |
| Main rivers: | None; many unnavigable streams |
| Highest elevation: | Tata Mai Lau, 9,721 ft (2,963 m) |
| Lowest elevation: | Sea level along the coasts |
| Form of government: | Republic |
| Head of state: | President |
| Head of government: | Prime minister |
| Administrative areas: | 13 districts |
| Legislature: | National Parliament with between 52 and 65 members serving five-year terms |
| Court system: | Supreme Court of Justice |
| Armed forces: | 1,300 troops |
| National holiday: | Independence Day - November 28 (1975) |
| Estimated 2010 population: | 1,192,000 |
| Population density: | 208 persons per mi² (80 per km²) |
| Population distribution: | 73% rural, 27% urban |
| Life expectancy in years: | Male, 62; female, 66 |
| Doctors per 1,000 people: | 0.1 |
| Birth rate per 1,000: | 40 |
| Death rate per 1,000: | 9 |
| Infant mortality: | 77 deaths per 1,000 live births |
| Age structure: | 0-14: 45%; 15-64: 52%; 65 and over: 3% |
| Internet users per 100 people: | 0.1 |
| Internet code: | .tl |
| Languages spoken: | Tetum (official), Portuguese (official), Indonesian, English |
| Religions: | Roman Catholic 98%, Muslim 1%, Protestant 1% |
| Currency: | United States dollar |
| Gross domestic product (GDP) in 2008: | $495 million U.S. |
| Real annual growth rate (2008): | 4.7% |
| GDP per capita (2008): | $498 U.S. |
| Goods exported: | Coffee, petroleum products |
| Goods imported: | Food, machinery, motor vehicles, petroleum products |
| Trading partners: | Australia, Indonesia, Singapore |

A mother and daughter wade in a rice paddy near Baukau on the northern coast of East Timor. Agriculture and fishing are important economic activities in East Timor.

The terrain of East Timor consists of coastal plains and a mountainous interior.

The man with whom he shared the Nobel Prize, Carlos Felipe Ximenes Belo, was born in a remote village in East Timor, where his family had lived for many generations. The Vatican consecrated him as bishop in 1988. As the leader of the church in East Timor, Belo became an active defender of the human rights of his people against the military forces of Indonesia.

## Independence from Indonesia

In a UN-sponsored referendum held in August 1999, the people of East Timor voted overwhelmingly for independence from Indonesia. Following the vote, armed pro-Indonesian militias, backed by some elements of the Indonesian military, began attacking and killing East Timorese citizens. Thousands of people were driven from their homes, and much of the East Timor capital of Dili was burned.

In mid-September, a UN-sanctioned multinational force began arriving in East Timor to try to restore peace to the region. In October, Indonesia's highest governmental body voted to accept the results of the referendum and to end Indonesia's claim to East Timor. The UN then set up an interim administration in East Timor to help prepare the region for full independence. In 2001, the people of East Timor elected an Assembly to create a constitution.

In April 2002, Xanana Gusmão was the winner of East Timor's first presidential election, with nearly 83 percent of the more than 378,500 votes cast. In 2007, Ramos-Horta was elected president and Gusmão became prime minister.

In February 2008, rebels attacked President Ramos-Horta and Prime Minister Gusmão in their homes during a failed coup attempt. Gusmão was not injured, but Ramos-Horta was critically wounded and hospitalized for two months. After the attacks, the government declared emergency rule to prevent further violence. The government lifted emergency rule in April.

# EASTER ISLAND

Remote Easter Island lies about 2,300 miles (3,700 kilometers) west of Chile. It is the easternmost island in Polynesia, more than 4,000 miles (6,400 kilometers) east of New Zealand. The stony, volcanic island covers 63 square miles (164 square kilometers). Its only fresh water comes from wells, pools, and crater lakes in the island's three extinct volcanoes.

## History and people

Scientists believe that Easter Island was settled between about A.D. 900 and 1200. The settlers were Polynesians who sailed to the remote island from islands to the west.

The early islanders created Easter Island's famous stone statues, which are called *moai* (pronounced *MOH eye*). A bloody war between two groups of Easter Islanders broke out about 1680. Over the next 150 years, the victors in that war and their descendants toppled the moai from their platforms—in most cases breaking the necks of the statues. Today, some of the moai have been restored to their original positions.

Easter Island's famous stone statues, the moai, were carved hundreds of years ago. More than 600 of the statues are scattered on the island, some rising 40 feet (12 meters). A number of moai have been restored to their original positions.

Extinct volcanoes dominate the landscape of Easter Island. Small lakes within the craters furnish much of the island's fresh water.

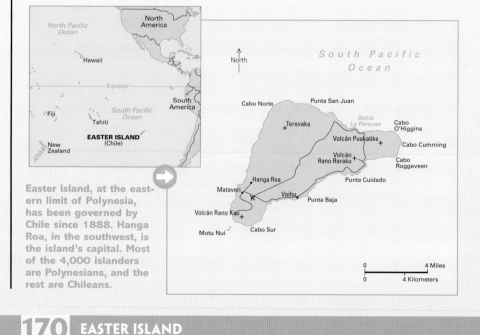

Easter Island, at the eastern limit of Polynesia, has been governed by Chile since 1888. Hanga Roa, in the southwest, is the island's capital. Most of the 4,000 islanders are Polynesians, and the rest are Chileans.

An Easter Islander displays his lobster catch. Many islanders fish and farm for a living, but tourism and the production of wool for export are the island's main industries.

Chile annexed the island in 1888 but neglected it for years, leasing about 90 percent of the land to private companies for sheep breeding. The native islanders retreated to a small reservation around Hanga Roa, the island's capital in the southwest. In the 1950's, Chile discovered the island's potential as a stopping point for international flights. As a result, Chile built a military airport, a school, a hospital, shops, and an agricultural institute on Easter Island. In 1965, the island was made a Chilean department, run by a governor and a native mayor.

Today, about 4,000 people live on the island. Most of them are Polynesians, and the rest are Chileans. Spanish, the language of Chile, is the island's official language, but the people also speak a Polynesian language called *Rapanui,* for the island's Polynesian name. Tourism and the production of wool for export are the main industries.

## The moai

Some scholars believe that the moai were intended to honor ancestors. Today, more than 600 of the statues are scattered on the island. Most are from 11 to 20 feet (3.4 to 6 meters) tall, while some rise an awe-inspiring 40 feet (12 meters) and weigh as much as 90 tons (82 metric tons). The most imposing collection stands where the statues were made, in what was once a quarry on the slopes of an extinct volcano called Rano Raraku in the eastern part of the island.

The islanders used stone hand picks to carve the statues from the lava rock of the extinct volcano. They set the statues on raised temple platforms called *ahu,* and they balanced huge red stone cylinders, like hats, on the heads of some of the statues. Even with modern technology, erecting such huge statues and balancing the cylinders on top of them would be a difficult task.

Jacob Roggeveen, a Dutch explorer, was the first European to see Easter Island. He discovered it on Easter Sunday, 1722, and gave the island its name. In 1862, slave ships from Peru arrived. Their crews kidnapped about 1,400 Easter Islanders and brought them to Peru to work on plantations. All but 100 of these islanders died in Peru, and the survivors were taken back to Easter Island in 1863.

During the voyage home, an additional 85 islanders died. The 15 survivors carried home the germs of smallpox and other diseases, which devastated the remaining population of Easter Island. During the early 1870's, many islanders left their homeland, and in 1877, only 110 people remained on Easter Island.

On the weekly market day known as the *feria* in the town of Otavalo, a woman peers out between the woven panels of cloth that are on display. The bold geometric designs and brilliant colors of the cloth are much like those used long ago by the woman's ancestors—the rich and powerful Inca. During the 1400's, the Inca conquered the Indian tribes who inhabited the region that is now Ecuador.

Inca rulers selected women of beauty and intelligence to be taught the art of weaving. These "chosen women" learned to weave wool and cotton thread into elaborately patterned cloth, which was used by the royal family and in religious ceremonies. The chosen women were also given some education and kept as "gifts" for nobles and other men who had performed a service for the rulers.

Although such customs disappeared when the Inca civilization fell, the skill of turning thread into cloth, of combining color and texture into a unique work of art, was never lost. Passed along from generation to generation, weaving is still practiced today in the Ecuadorean highlands, much as it was centuries ago.

Spanish conquistadors vanquished the Inca in 1534. Most of the Spaniards settled in the Andes Highlands, where they made Quito their capital. They built many beautiful churches and public buildings in the city, and Quito became a great center of religious art. The Spanish settlers also created large farms and estates called *haciendas,* and they forced the conquered Indians to work on them.

During the 1800's, when Ecuador was a Spanish colony, the French emperor Napoleon invaded and conquered Spain. The rulers of the Spanish colonies in South America, including Ecuador, took advantage of Spain's weakness to demand independence.

In 1822, General Antonio José de Sucre defeated the Spaniards, which ended Spanish rule in Ecuador. Ecuador then joined newly independent Colombia and Venezuela in the confederation of Gran Colombia. In 1830, Ecuador left the confederation and became an independent country.

# ECUADOR

Rival leaders fought for power in the new nation. Presidents, dictators, and *juntas* (groups of rulers) rose and fell. In their hunger for control, most showed no concern for the rights and needs of the people. One exception was Gabriel García Moreno, who became president in 1861. His government helped develop farming and industry and built roads and railroads. But he was assassinated in 1875, and his plans and policies were not continued.

During a particularly unstable period between 1925 and 1948, Ecuador had 22 presidents and heads of state, and none of them served a complete term. In 1963, armed forces overthrew the civilian government. Civilian and military rulers alternated in control of the government between 1963 and the civilian elections of 1979, when elections began being held regularly.

In early 1997, Ecuador's legislature removed President Abdala Bucaram Ortiz from office, claiming he was mentally unfit to serve, appointing an interim president until elections could be held. In 1998, voters elected Jamil Mahuad president.

Ecuador faced economic difficulties in 1999 and early 2000. The nation failed to pay back loans from several international lending organizations, and its currency dropped sharply in value. In January 2000, a coalition of military officers and Indian groups forced Mahuad from office.

In December 2004, Ecuador's Congress voted to dismiss and replace almost all of the judges on the Supreme Court. President Lucio Gutierrez Borbua, elected in 2002, had claimed that the judges favored his political opponents. After the new Supreme Court overturned corruption charges against former President Bucaram Ortiz, an ally of Gutierrez, protesters organized demonstrations demanding Gutierrez's resignation. Congress then voted Gutierrez out of office.

In November 2006, voters elected leftist candidate Rafael Correa president. Correa called for the country to reform its constitution. In September 2008, Ecuadoreans approved a new constitution that gives more powers to the president. Correa was reelected in 2009.

# ECUADOR TODAY

Rich in natural beauty, Ecuador is a country of magnificent landscapes as yet unspoiled by mass tourism. The country's highlands, reached by air or railroad, are breathtaking. Delightful beaches line its Pacific coast, while Quito's churches and monasteries offer an abundance of exquisite colonial paintings and sculptures. The offshore Galapagos Islands, about 600 miles (970 kilometers) west of Ecuador, are a living museum of some of the world's most unusual wildlife.

Yet the beauty of this land, with its fiery volcanic peaks, thick tropical forests, and vast mountain plateaus, is a stark contrast to the bleak and often difficult life of most present-day Ecuadoreans. Many Indians, *mestizos* (people of mixed Indian and European ancestry), and blacks—who together make up about 93 percent of Ecuador's population—live in terrible poverty. Often, the poor people in rural areas suffer from malnutrition and *dysentery,* an intestinal disease.

Many mestizos live in wooden homes with thatched roofs and work as day laborers on large banana or cacao plantations. Most of the blacks live in the northern part of the coastal plain, and many of them fish for a living. Some Ecuadoreans live in the forests, cultivating small plots of land and moving on as the soil wears out.

Most of Ecuador's Indians live in Andean villages and have little contact with the rest of the people. Some of them work on haciendas, growing crops and herding livestock. Others are shepherds in pastures high in the mountains.

In sharp contrast to these remote areas is the bustling, urban elegance of Quito, Ecuador's capital. From its glass-and-concrete skyline to its Baroque churches, from its teeming Indian markets to its Spanish-style palaces, the architecture of Quito reflects its history.

The residents of Quito and other Andean cities include many whites of European ancestry. In Ecuador, where they make up about 7 percent of the popula-

## FACTS

| | |
|---|---|
| Official name: | Republica del Ecuador (Republic of Ecuador) |
| Capital: | Quito |
| Terrain: | Coastal plain (costa), inter-Andean central highlands (sierra), and flat to rolling eastern jungle (oriente) |
| Area: | 109,484 mi² (283,561 km²) |
| Climate: | Tropical along coast, becoming cooler inland at higher elevations; tropical in Amazonian jungle lowlands |
| Main rivers: | Esmeraldas, Daule, Napo, Curaray |
| Highest elevation: | Chimborazo Volcano, 20,702 ft (6,310 m) |
| Lowest elevation: | Pacific Ocean, sea level |
| Form of government: | Republic |
| Head of state: | President |
| Head of government: | President |
| Administrative areas: | 24 provincias (provinces) |
| Legislature: | Congreso Nacional (National Congress) with 100 members serving four-year terms |
| Court system: | Corte Suprema (Supreme Court) |
| Armed forces: | 58,000 troops |
| National holiday: | Independence Day - August 10 (1809) |
| Estimated 2010 population: | 14,012,000 |
| Population density: | 128 persons per mi² (49 per km²) |
| Population distribution: | 63% urban, 37% rural |
| Life expectancy in years: | Male, 72; female, 78 |
| Doctors per 1,000 people: | 1.5 |
| Birth rate per 1,000: | 21 |
| Death rate per 1,000: | 5 |
| Infant mortality: | 21 deaths per 1,000 live births |
| Age structure: | 0-14: 32%; 15-64: 62%; 65 and over: 6% |
| Internet users per 100 people: | 10 |
| Internet code: | .ec |
| Languages spoken: | Spanish (official), Amerindian languages (especially Quechua) |
| Religions: | Roman Catholic 95%, other 5% |
| Currency: | United States dollar |
| Gross domestic product (GDP) in 2008: | $53.27 billion U.S. |
| Real annual growth rate (2008): | 5.9% |
| GDP per capita (2008): | $3,851 U.S. |
| Goods exported: | Bananas, cocoa, coffee, crude oil, cut flowers, shrimp |
| Goods imported: | Chemicals, machinery, motor vehicles, petroleum products |
| Trading partners: | China, Colombia, Peru, United States, Venezuela |

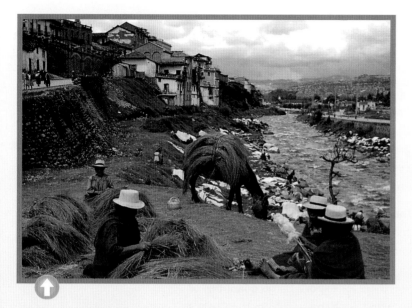

Workers in Cuenca bale sisal, a fiber used to manufacture rope and twine. They wear Panama hats, one of Ecuador's best-known exports.

tion, white Europeans are the wealthiest and most powerful group. Some are leaders of business and industry, and some own the land worked by the Indian and mestizo farmers in the rural areas.

These landlords sometimes hire managers to run their large haciendas. Unlike the Andean villagers, who build simple houses of adobe to protect themselves from the mountain cold, many wealthy white people live in comfortable, stylish city residences. Some also have weekend country residences.

The uneven distribution of wealth is not unique to this small South American country. Like its neighbors to the north, east, and south, Ecuador struggles with this and other economic problems. However, the government is working to improve the living standards of its people. Several programs have been set up to provide homes, improve medical care, and promote literacy.

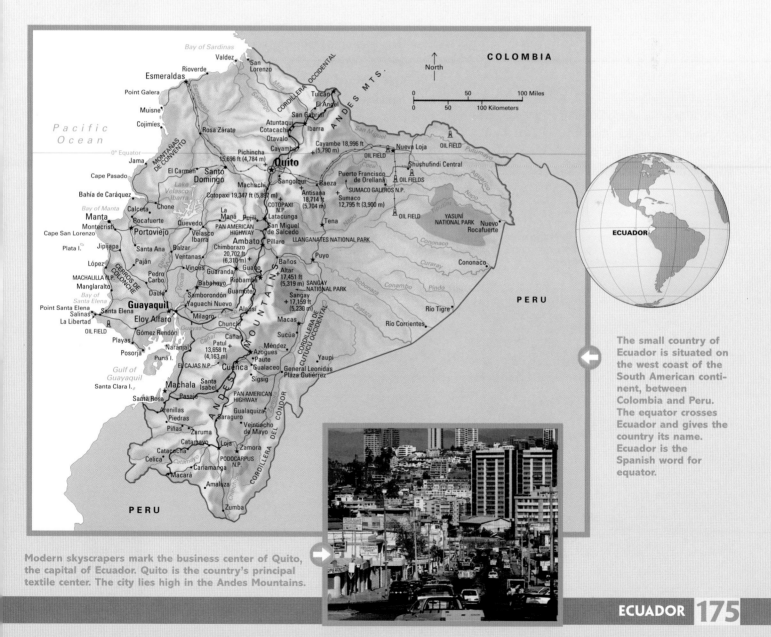

The small country of Ecuador is situated on the west coast of the South American continent, between Colombia and Peru. The equator crosses Ecuador and gives the country its name. Ecuador is the Spanish word for equator.

Modern skyscrapers mark the business center of Quito, the capital of Ecuador. Quito is the country's principal textile center. The city lies high in the Andes Mountains.

# ENVIRONMENT

Ecuador has a rugged and varied landscape, ranging from the wet, swampy forests of the Eastern Lowland areas to the snow-capped volcanic peaks of the Andes Mountains. Although the equator crosses Ecuador in the northern region of the country, the climate is not always hot because much of the country lies at higher altitudes, where the air is cooler.

Geographers divide the mainland of Ecuador into three land regions: the Coastal Lowland, also known as the Costa; the Andes Highland, also called the Sierra; and the Eastern Lowland, or Oriente. The Galapagos Islands, an offshore province and Ecuador's fourth land region, lie about 600 miles (970 kilometers) off the mainland in the Pacific Ocean.

## Land regions

The Coastal Lowland, or Costa, is a large, flat plain that extends along Ecuador's Pacific coast and covers about a fourth of the country. It was formed by mud and sand sediments carried down the mountains by rivers and deposited along the shore. In the north, the Costa is wet and swampy. Dense tropical jungles cover the land, stretching all the way up the Andean slopes to 8,000 feet (2,400 meters) in some places. In the south, near Peru, the Costa is a desert.

The warm, humid conditions and fertile soil of the northern Costa make it a productive farming region. Abundant crops of bananas, cacao, coffee, oranges, rice, and sugar cane are grown there for export. Large quantities of balsa wood are harvested from the Costa's forests, making Ecuador the world's largest producer of this lightweight wood.

These products are shipped from Guayaquil, Ecuador's leading commercial center and seaport and its largest city. The area surrounding Guayaquil has two seasons: a hot, rainy period from January to May and a cooler period from June to December. During the cooler period, sea breezes from the Peru Current (also called the Humboldt Current) help ease the equatorial heat.

Shepherds lead their flock to pasture on the lower slopes of Chimborazo volcano the highest of about 30 Andes mountains that form an "avenue of volcanoes." Chimborazo's tall peak is covered with snow all the time.

A group of Andean shepherds wrapped in blankets to shield them against the cold winds, rest for a moment from their work. Many sheep in Ecuador are grazed at altitudes over 9,000 feet (2,700 meters).

Bay of Sardinas

Esmeraldas
Point Galera

Esmeraldas

ANDES
HIGHLAND

Pichincha
15,696 ft (4,784 m)

Cayambe 18,996 ft (5,790 m)

Equator

Cape Pasado

Lake
Velasco
Ibarra

Quito

Sumaco
12,795 ft (3,900 m)

Bay of Manta

Manta

Cotopaxi
19,347 ft (5,897 m)

Antisana
18,714 ft (5,704 m)

Cape San Lorenzo

COASTAL
LOWLAND

Chimborazo
20,702 ft (6,310 m)

EASTERN
LOWLAND

Altar 17,451 ft (5,319 m)

Bay of
Santa Elena

Point Santa Elena

Guayaquil

Paute
13,650 ft
(4,162 m)

Sangay
17,159 ft
(5,230 m)

Puná I.

Cuenca

Gulf of Guayaquil

Machala

ANDES
HIGHLAND

Ecuador's landscape rises dramatically from sea level along the coast to mountain peaks in the central Andean Highland that stand more than 20,000 feet- (6,100 meters) high.

The Andean Highland, or Sierra, makes up another fourth of Ecuador's land area, with two parallel ranges extending from north to south along the entire length of the country. Between them lie a series of high plateaus and basins, where Andean farmers grow beans, corn, and potatoes, much as their ancestors did thousands of years ago. Cattle graze in the highland valleys. There are few native trees in the highlands, though the eucalyptus, introduced in the 1860's, is now widely grown in the area. Temperatures on the Sierra plateaus are generally springlike, with colder weather at higher altitudes.

The heavily forested Eastern Lowland, or Oriente, forms part of the Amazon Basin and makes up almost half of Ecuador's land area. Temperatures are high, and rain falls all year long. In the past, the Oriente was virtually undeveloped. However, since large petroleum deposits were found there, the area now contains a number of oil fields.

## Environmental problems

Ecuador's tropical rain forests have a wide variety of wildlife, including deer, jaguars, monkeys, ocelots, tapirs, and many species of birds. To protect its wildlife, the Ecuadorean government established a broad wildlife-protection program in 1970.

Despite these efforts, extensive clearing of the forests for farmland and oil field development continue to threaten Ecuador's native species. For instance, the black caiman, once an inhabitant of Ecuador, is now extinct.

An Indian woman in the Andean Highland does the family wash in the cool waters of a mountain stream. Living conditions are often quite simple in Ecuador's rural areas, and many people use old-fashioned equipment and tools to farm the land.

Mestizo women take shelter from a rain shower in a grove of sugar cane in the northern Costa. Crops for export are grown mainly in the lowlands. Most products grown in the highlands are used by the local people.

An Otavalo Indian woman prepares wool before threading her loom. The Otavalos have long been skilled weavers. They often travel as far as Colombia, Venezuela, Brazil, and the United States to sell their ponchos, blankets, and textiles.

# GALAPAGOS ISLANDS

Far out into the Pacific, about 600 miles (970 kilometers) west of the Ecuadorean mainland, lies a chain of islands whose unusual animal inhabitants have fascinated scientists and voyagers for centuries. They are the Galapagos Islands, named for the giant tortoises that live there.

The islands are made up of volcanic peaks and cover an area of 3,029 square miles (7,844 square kilometers). Once called the Enchanted Isles, the Galapagos may have been originally settled by Peruvian Indians. During the 1600's and 1700's, pirates and buccaneers used the islands as hideouts from which they launched raids on Spanish ships and the coastal towns of South America.

The human population of the Galapagos has increased greatly since the 1980's. Between the censuses of 1982 and 2001, the population more than tripled, growing from about 6,000 to nearly 19,000. In addition, thousands of tourists visit each year.

The island is known for its exotic birds and other animals. For example, marine lizards that measure 4 feet (1.2 meters) long slither over the rocks, while 500-pound (230-kilogram) tortoises plod along the ground, often with Galapagos hawks perched on their backs.

**Pinnacle Rock, on the island of Bartholomé, is a favorite nesting place of the Galapagos penguins. Because of its volcanic origin, the land surface of the islands is made up of basaltic lava. Visitors often compare it to the barren surface of the moon. There are almost no tall trees or other plants because the islands have no deep soil where trees can take root. One exception is the prickly pear cactus, which towers up to 30 feet (9 meters) in the arid lowlands of some islands.**

## FINCHES

**The Galapagos Islands became famous all over the world after Charles Darwin published his book *The Origin of Species* (1859). Darwin was a British naturalist who traveled to the islands on a scientific expedition aboard the H.M.S. *Beagle*.**

**Darwin studied the finches of the Galapagos Islands. He noted that although the finches were basically similar, a number of different types had evolved. They range from the ground finch, with its thick bill for crunching hard seed, to the tiny warbler finch, which feeds on insects. Darwin realized that this one species had filled all the roles that, in the United Kingdom, were filled by a large variety of birds. He concluded that the Galapagos finches had all descended from a single ancestor finch that had flown to the islands from the South American mainland.**

**Darwin's findings led him to develop his theories on evolution and a process called natural selection, or survival of the fittest. *The Origin of Species*, which discusses these theories, sparked heated debate among the biologists and religious leaders of the time.**

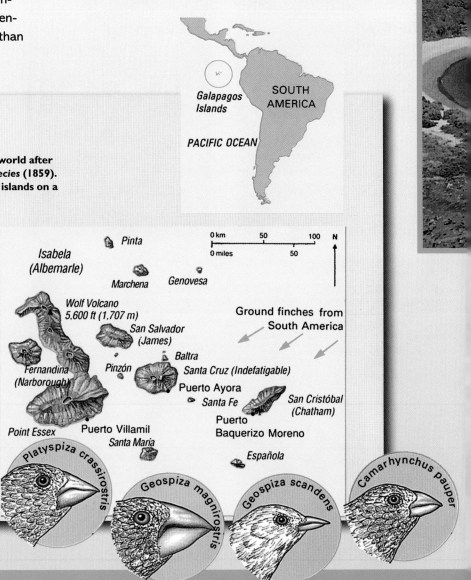

## Wonders of the natural world

The animals that lived on the islands before people arrived either flew in, swam, or rode in on floating vegetation or on the backs of other animals. The only mammals among these animals were two species of bats, several species of rats, and sea lions and fur seals.

Of the islands' many birds, the flightless cormorant is perhaps the most remarkable. It has stunted wings and walks on land with an upright waddle. The cormorant moves very well in water, however, enabling it to compete with penguins for food.

The marine iguanas of Galapagos are the only lizards in the world that live in the sea. They often dive deep into the water to feed on marine algae and shellfish, using their large, clawed feet to dig into the ocean bottom.

The marine iguanas can drink seawater, excreting the excess salt through glands in their noses.

The giant, plant-eating tortoises of the Galapagos graze mainly on the native prickly pear cactus.

# EGYPT

Perhaps nothing symbolizes Egypt better than the great pyramids at Giza. Built more than 4,500 years ago, the pyramids are the magnificent remains of a great civilization headed by strong rulers called *pharaohs*. Among the outstanding contributions to civilization made by ancient Egyptians are a 365-day calendar, a form of picture writing called *hieroglyphics,* and certain basic forms of arithmetic. Egyptians also built great cities in which skilled architects, doctors, engineers, painters, and sculptors worked.

Egypt is no longer a wealthy world power, but much of what was true for ancient Egypt holds true for modern Egypt as well. The ancient Egyptians created the world's first national government, and a central government has controlled the nation almost continuously ever since. The ancient Egyptians' religion influenced their everyday life as well as their political life. They believed their pharaoh was a god, and priests held great power. Today in Egypt, the religion of Islam influences family life, social relationships, business activities, and government affairs.

Just as in ancient Egypt, Egyptians still cluster near the Nile River—on its fan-shaped delta and the two narrow strips of fertile land that lie along the riverbanks. Without the precious waters of the Nile, all of Egypt would be a desert.

West of the Nile Valley lies the Libyan Desert, part of the huge Sahara that stretches across northern Africa. East of the Nile lies the Arabian Desert, also part of the Sahara. Even the triangular Sinai Peninsula to the east, across the Gulf of Suez and the Suez Canal, is arid and desolate.

Egypt's greatest problem today may be overpopulation. More than 81 million people are now crowded into the Nile's narrow valley and delta. The cities of Egypt, in addition to having such typical urban problems as housing shortages and traffic congestion, are overflowing. Many city people live in extreme poverty, while others enjoy all the conveniences of modern life.

Meanwhile, village farmers live much as their ancestors did centuries ago. And the people of Egypt still rely heavily on the crops produced on the fertile Nile land, much as the ancient Egyptians did thousands of years ago. Whether they are city dwellers or rural villagers, whether rich or poor, Egypt's people all share common bonds—the beliefs and traditions of Islam and a rich cultural history.

181

# EGYPT TODAY

Egypt is a Middle Eastern country tucked into the northeast corner of Africa. Once a rich and powerful ancient kingdom, it became a modern, independent republic in 1953. Since then, Egypt has played a leading role in the Middle East.

According to its 1971 Constitution, the country is a democratic and socialist society. A military government took control of Egypt in 2011 and replaced the 1971 Constitution with a temporary charter. The charter was to remain in effect until a new constitution could be written and approved by a direct vote of the people.

Under the old Constitution, Egypt's president had great power in all levels of government. The People's Assembly did little more than approve the president's policies.

The National Democratic Party was Egypt's largest political party until an Egyptian court dissolved it in 2011. Opposition parties began to be allowed to participate in general elections in 2005. The most influential opposition group was the Muslim Brotherhood, which favored a government based on Islamic principles and law.

In the 1990's and early 2000's, violence erupted between the government and Islamic fundamentalist extremists seeking to topple the government. The government responded with mass arrests and executions.

In January 2011, antigovernment protests erupted across Egypt. Protesters called for an end to the presidency of Hosni Mubarak, who had been in power since 1981. More than 800 demonstrators were killed in the uprising. Mubarak stepped down in February 2011, and the Egyptian military took control of the government. Mubarak was tried and sentenced to life in prison for the killings.

## FACTS

| | |
|---|---|
| Official name: | Jumhuriyat Misr al-Arabiyah (Arab Republic of Egypt) |
| Capital: | Cairo |
| Terrain: | Vast desert plateau interrupted by Nile valley and delta |
| Area: | 386,662 mi² (1,001,450 km²) |
| Climate: | Desert; hot, dry summers with moderate winters |
| Main river: | Nile |
| Highest elevation: | Jabal Katrinah, 8,651 ft (2,637 m) |
| Lowest elevation: | Qattara Depression, 436 ft (133 m) below sea level |
| Form of government: | Transitional |
| Head of state: | President |
| Head of government: | Prime minister |
| Administrative areas: | 29 muhafazat (governorates) |
| Legislature: | Majlis al-Sha'b (People's Assembly) with up to 454 members serving five-year terms and the Majlis al-Shura (Advisory Council) with 264 members serving six-year terms |
| Court system: | Supreme Constitutional Court |
| Armed forces: | 468,500 troops |
| National holiday: | Revolution Day - July 23 (1952) |
| Estimated 2010 population: | 81,495,000 |
| Population density: | 211 persons per mi² (81 per km²) |
| Population distribution: | 57% rural, 43% urban |
| Life expectancy in years: | Male, 70; female, 74 |
| Doctors per 1,000 people: | 2.4 |
| Birth rate per 1,000: | 24 |
| Death rate per 1,000: | 6 |
| Infant mortality: | 30 deaths per 1,000 live births |
| Age structure: | 0-14: 33%; 15-64: 62%; 65 and over: 5% |
| Internet users per 100 people: | 15 |
| Internet code: | .eg |
| Languages spoken: | Arabic (official), English, French |
| Religions: | Muslim (mostly Sunni) 90%, Coptic Christian 9%, other Christian 1% |
| Currency: | Egyptian pound |
| Gross domestic product (GDP) in 2008: | $162.16 billion U.S. |
| Real annual growth rate (2008): | 6.9% |
| GDP per capita (2008): | $2,099 U.S. |
| Goods exported: | Clothing; crude oil and petroleum products; food, especially fruits, vegetables, and rice; iron and steel |
| Goods imported: | Food, especially corn and wheat; machinery; petroleum products; transportation equipment |
| Trading partners: | Germany, Japan, Saudi Arabia, United States |

Egypt is important in global transportation, in the Arab world, and in Middle East politics because of its strategic location. In 1956, Egypt took control of the Suez Canal, which links the Mediterranean Sea with the Red Sea. In 1967 and 1973, Egypt joined other Arab nations in fighting Israel, but in 1979 Egypt and Israel signed a major peace agreement.

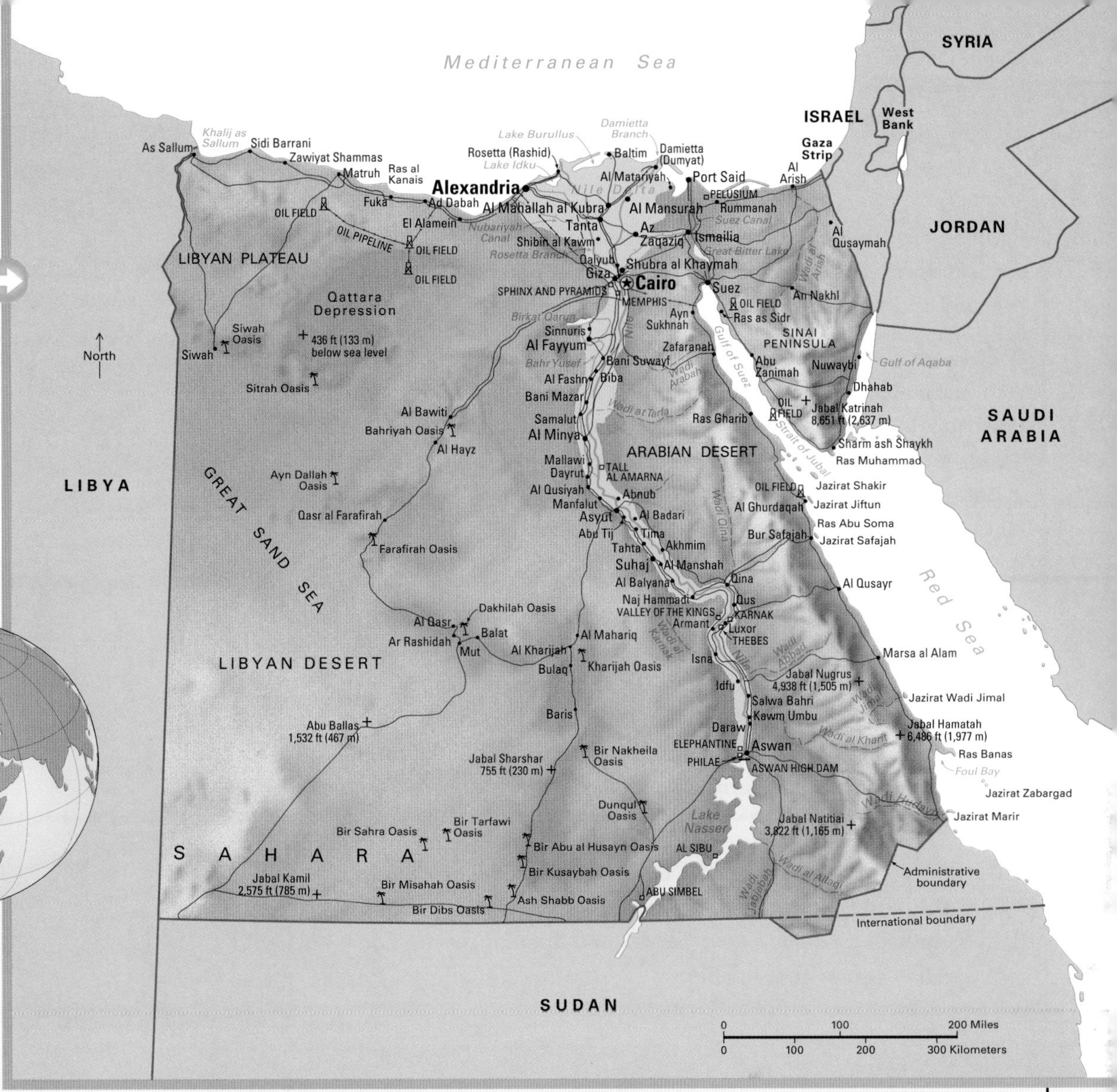

Mediterranean Sea
SYRIA
ISRAEL
West Bank
Gaza Strip
JORDAN
SAUDI ARABIA
LIBYA
SUDAN
Red Sea
Gulf of Suez
Gulf of Aqaba
Strait of Jubal
North
As Sallum
Khalij as Sallum
Sidi Barrani
Zawiyat Shammas
Matruh
Ras al Kanais
Alexandria
Rosetta (Rashid)
Lake Idku
Lake Burullus
Damietta Branch
Baltim
Damietta (Dumyat)
Port Said
Al Arish
Al Matariyah
PELUSIUM
Rummanah
Al Mansurah
Al Mahallah al Kubra
Nile Delta
Tanta
Az Zaqaziq
Ismailia
Suez Canal
Great Bitter Lake
Shibin al Kawm
Qalyub
Shubra al Khaymah
Giza
Cairo
SPHINX AND PYRAMIDS
MEMPHIS
Suez
Fuka
Ad Dabah
El Alamein
Nubariyah Canal
Rosetta Branch
OIL FIELD
OIL PIPELINE
LIBYAN PLATEAU
Qattara Depression
436 ft (133 m) below sea level
Siwah Oasis
Siwah
Sitrah Oasis
Birkat Qarun
Sinnuris
Al Fayyum
Bahr Yusef
Ayn Sukhnah
Zafaranah
Bani Suwayf
Al Fashn
Biba
Bani Mazar
Samalut
Al Minya
Al Bawiti
Bahriyah Oasis
Al Hayz
Al Qusaymah
An Nakhl
Ras as Sidr
SINAI PENINSULA
Abu Zanimah
Nuwaybi
Dhahab
Jabal Katrinah 8,651 ft (2,637 m)
Sharm ash Shaykh
Ras Muhammad
Ras Gharib
ARABIAN DESERT
Mallawi
Dayrut
TALL AL AMARNA
Al Qusiyah
Manfalut
Abnub
Asyut
Abu Tij
Al Badari
Tima
Tahta
Akhmim
Suhaj
Al Manshah
Al Balyana
Naj Hammadi
VALLEY OF THE KINGS
Qina
Qus
KARNAK
Armant
Luxor
THEBES
Isna
Idfu
Salwa Bahri
Kawm Umbu
Daraw
ELEPHANTINE
Aswan
PHILAE
ASWAN HIGH DAM
Lake Nasser
AL SIBU
ABU SIMBEL
Jazirat Shakir
Jazirat Jiftun
Al Ghurdaqah
Ras Abu Soma
Bur Safajah
Jazirat Safajah
Al Qusayr
Marsa al Alam
Jabal Nugrus 4,938 ft (1,505 m)
Jazirat Wadi Jimal
Jabal Hamatah 6,486 ft (1,977 m)
Ras Banas
Foul Bay
Jazirat Zabargad
Jazirat Marir
Jabal Natitiai 3,822 ft (1,165 m)
Administrative boundary
International boundary
Ayn Dallah Oasis
GREAT SAND SEA
Qasr al Farafirah
Farafirah Oasis
Dakhilah Oasis
Al Qasr
Balat
Ar Rashidah
Mut
LIBYAN DESERT
Al Kharijah
Al Maharig
Bulaq
Kharijah Oasis
Baris
Abu Ballas 1,532 ft (467 m)
Jabal Sharshar 755 ft (230 m)
Bir Nakheila Oasis
Dunqul Oasis
Bir Sahra Oasis
Bir Tarfawi Oasis
SAHARA
Bir Abu al Husayn Oasis
Bir Kusaybah Oasis
Jabal Kamil 2,575 ft (785 m)
Bir Misahah Oasis
Ash Shabb Oasis
Bir Dibs Oasis
0 100 200 Miles
0 100 200 300 Kilometers


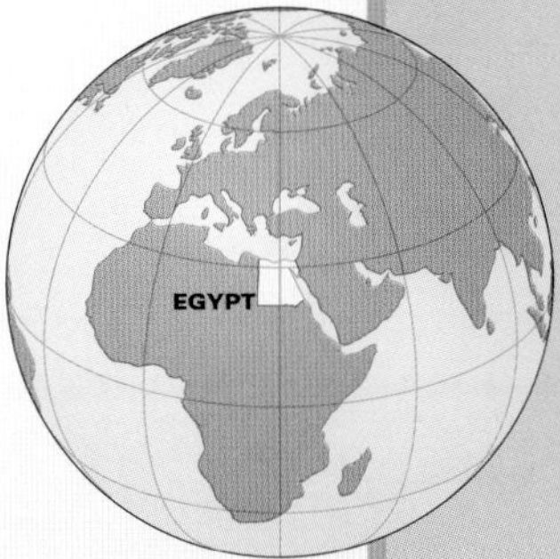

EGYPT


In late 2011, amidst demonstrations against military rule, Egyptians elected a new parliament. The Muslim Brotherhood emerged as the most powerful political party in the country. A constitutional assembly began to draft a new constitution. In 2012, Egypt held its first free presidential election.

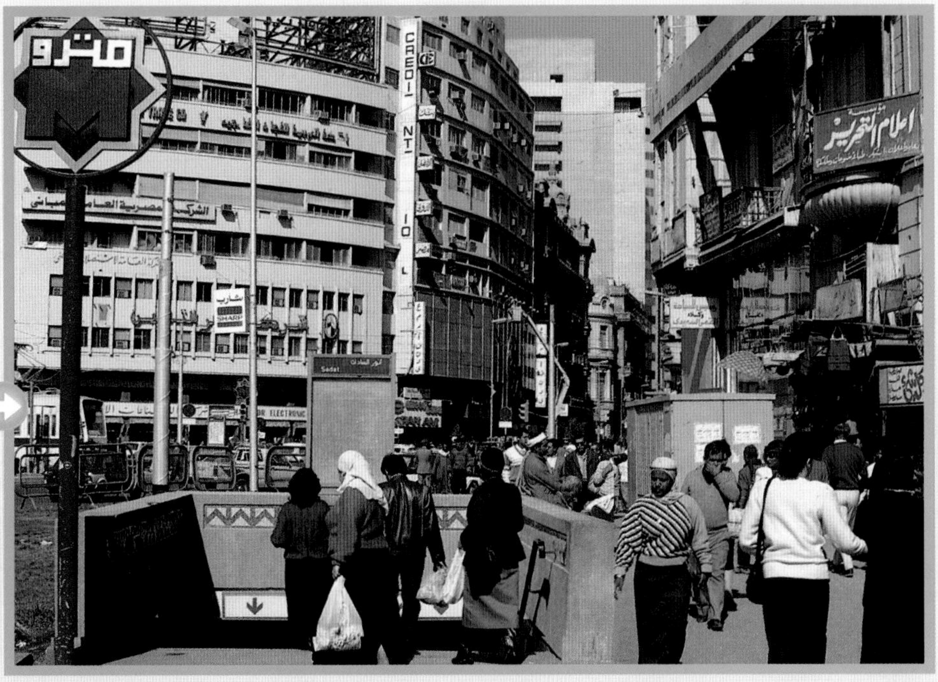

Sadat
SHARP


On a busy Cairo street, people enter the city's Metro, an underground railway system that opened in 1987. Transportation in Egypt takes a variety of forms, as buses, cars, trucks, and motorcycles share the roads with bicycles, donkeys, and carts.

# AGRICULTURE

Farming has been central to the Egyptian economy for thousands of years. The mighty Nile River has nourished the land and the people of Egypt since ancient times.

Every year in July, the Nile started to flood, swelling with rain water that had fallen in central Africa and flowed downstream. Egyptian farmers trapped the floodwaters in basins on their fields. When the river went down, usually in September, it left a strip of rich, black soil about 6 miles (10 kilometers) wide on each bank. The farmers then plowed and seeded the fertile soil.

In the 1800's, the Egyptians began to replace their seasonal system of basin irrigation with a year-round system of dams, canals, and reservoirs. The changeover was completed in 1968, when the Aswan High Dam began operating. Today, nearly all of Egypt's farmland can be irrigated continuously, and farmers can plant crops the year around.

Cotton is the nation's most valuable crop, and Egypt is one of the world's leading cotton producers. Egyptian farmers grow high-quality, *long-staple* (long-fibered) cotton, which is strong and durable.

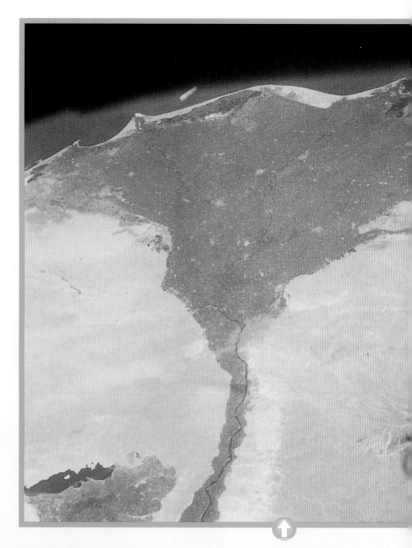

A satellite photo of the Nile Delta shows its fanlike shape, which some people think resembles an Egyptian lotus flower. The river forms a narrow life-giving strip in the desert, like the stem of the lotus, then fans out at its delta, like the flower itself.

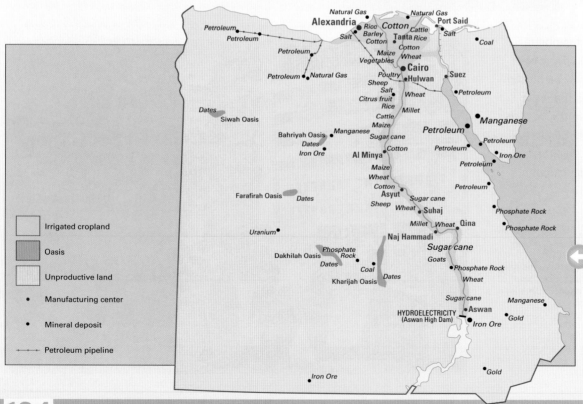

Irrigated cropland

Oasis

Unproductive land

• Manufacturing center

• Mineral deposit

—•— Petroleum pipeline

Egypt's farmland lies mostly along the Nile River and its delta. About 70 percent of Egypt's farms cover 1 acre (0.4 hectare) or less, and only about 6-1/2 million acres (2.6 million hectares) are cultivated—less than 3 percent of the country's total land area.

## IRRIGATION

The traditional Egyptian method of irrigation was basin irrigation, top. Farmers built canals and small dikes on their land to create basins that would catch the floodwaters of the Nile. When the floodwaters eventually went down, rich black silt remained. As farmers learned to clear more land, center, they increased both the number of crops they could raise and the length of the growing season. Devices for lifting water buckets called shadoofs or water wheels called sakiehs were used to bring water to fields on higher ground. Since the 1800's, modern irrigation—a system of dams, reservoirs, pumps, and canals—has been constructed, bottom. But now, modern farmers along the Nile must use chemical fertilizers because the Aswan High Dam holds back the fertile silt.

Using a primitive water wheel, an Egyptian *fellah* (farmer) draws water to irrigate his field. Irrigation is a necessity in most parts of Egypt, where little rain falls.

Egyptian rural peasants—called *fellahin*—have farmed small plots of land or tended animals for hundreds of years. However, many fellahin did not own the land they farmed. In the 1950's and 1960's, the Egyptian government tried to help the fellahin by passing land-reform measures that limited the size of farm estates and prevented large, wealthy landowners from controlling so much valuable farmland. But for many fellahin, little has changed. They still rent land or work as laborers in the fields of prosperous landowners.

The government requires farmers to use some of their land for growing the food crops that feed Egypt's increasing population, rather than cash crops such as cotton. Farmers grow corn, oranges, potatoes, rice, sugar cane, tomatoes, and wheat. Dates are grown mainly in desert oases. Farmers and herders raise goats and sheep for meat, milk, and wool. Many farmers also raise chickens for meat and eggs.

# PEOPLE

More than 81 million people live in Egypt, and almost all of them are crowded into the narrow Nile Valley and the Suez Canal region. And in these areas, almost half of all Egyptians are further crowded into the cities.

Over hundreds of years, many people from other lands have invaded Egypt and married native Egyptians. As a result, present-day Egyptians can trace their ancestry not only to ancient Egyptians, but also to Arabs, Ethiopians, Persians, and Turks, as well as Greeks, Romans, and other Europeans.

However, most Egyptians consider themselves Arabs. Arab Muslims surged into Egypt and conquered the land in the A.D. 600's. The Egyptian people gradually adopted the Arab language, and most converted from the Coptic Christian religion to Islam.

Arabic is now the official language of Egypt, though people in different regions speak different *dialects* (forms) of Arabic. The dialect spoken in Cairo is the one most widely used today. Many educated Egyptians also speak English or French.

Islam is the official religion of Egypt. About 90 percent of Egypt's people are Muslims, almost all of whom follow the Sunni branch of Islam. Under the law, Coptic Christians

Cairo commuters squeeze on and off their train at a bustling Metro stop. People in Cairo use a variety of forms of transportation, including buses, cars, motorcycles, and trains.

A Cairo coffee house provides a pleasant place for city dwellers to relax and chat with friends.

A Nubian woman in Aswan balances a bucket on her head in the traditional method of carrying a burden.

and other religious minorities may worship freely, but some Muslim groups have committed acts of violence against Coptic communities.

Egypt has several ethnic minority groups. The Bedouin are Arab nomads who traditionally lived in the desert, though most are now settled farmers. They speak Bedouin dialects. The Nubians originally lived in southern Egypt and farmed along the Nile. In 1968, when their land was flooded by the Aswan High Dam, the government relocated the Nubians to Kawm Umbu, a town about 30 miles (50 kilometers) north of Aswan. However, like other rural Egyptians, many Nubians have moved to the cities to look for work.

Life in the cities differs greatly from life in the countryside. And there are great differences in urban lifestyles because there are such extremes of wealth and poverty in Egypt's cities.

Some city people live in pleasant residential areas. Many well-to-do Egyptians wear clothing similar to that worn in North America and Europe. They buy large quantities of meat and imported fruits and vegetables. However, many more urban dwellers live in sprawling slums—in crowded, run-down apartments, in makeshift huts, or even on the roofs of buildings. The city poor eat a simple diet of bread and broad beans called *ful* or *fool*. Government-run stores sell meat, cheese, eggs, and other foods at controlled prices, but supplies often run out.

Until recently, most rural Egyptians lived in small, mud-brick huts with thatched straw roofs. Today, many of their homes are built of fired bricks or concrete, and they are larger and more comfortable than in the past. Televisions and other modern amenities are widely available.

Each member of a fellahin family has certain duties. The husband organizes the planting, weeding, and harvesting, while the wife cooks, carries water, and helps in the field. Children tend animals and help carry water. Fellahin men traditionally wear pants and a long, shirtlike garment called a *galabiyah,* and the women wear long flowing gowns. The fellahin diet is similar to that of the urban poor.

Islam provides a bond between the people, and rural Egyptians have a strong sense of community. Religion influences many aspects of Egyptian life. Muslims are expected to pray five times a day, give money to the poor, and fast. Some Egyptians dress according to modern Islamic customs. The men wear long, light-colored gowns and skullcaps, and the women wear robes and keep their faces veiled in public.

Traditionally, family roles in Egypt have been rigid, and women have had low status. However, since the 1960's, many more Egyptian women attend school or work outside the home.

Bedouin women spin wool into yarn, practicing a craft that was developed about 4000 B.C.

# AL-AZHAR UNIVERSITY

Egypt's influential position as the traditional center of Arab learning and culture has become even more important since the 1970's, when a wave of strong religious feeling developed among the various segments of many Muslim nations. Much of Egypt's leadership in the Muslim world stems from a historic source—Cairo's Al-Azhar University. Al-Azhar is one of the oldest universities in the world and the most influential religious school in Islam.

Al-Azhar was founded about A.D. 970 by the Fatimids, a band of Shi'ah Muslims who invaded Egypt from the west. The Fatimid *caliphs,* or rulers, claimed they were descended from Fatimah, the daughter of the prophet Muhammad, who founded Islam. The Fatimids established a capital city called *Al Qahirah*—perhaps because the planet Mars (*al-Qahir* in Arabic) was rising in the sky when the city was founded. The name *Cairo* came from *Al Qahirah.*

The Fatimids erected beautiful buildings and a great library in Cairo, which soon became one of the most important cities of the Arab world. The Fatimids also built Al-Azhar Mosque, a Muslim place of worship that quickly became the center of Fatimid culture and religion. Eventually, Al-Azhar became a university, a center for the study of Islamic religion and law that attracted students of Islam from many countries.

For hundreds of years, the leading scholars of Al-Azhar had great power and influence in Egypt. Like the priests of ancient Egypt, these religious scholars, called *ulama,* sometimes even controlled Egyptian rulers.

For centuries, almost all of Egypt's educational, legal, social welfare, and health affairs were in the hands of these religious leaders. However, during the 1800's and 1900's, the government brought religious institutions such as Al-Azhar under closer control and limited the role of the ulama in public life.

In 1952, a group of Egyptian military officers seized power and brought about even more dramatic changes. The government took over the appointment of officials to mosques and religious schools. The ulama then came under close government supervision and financial control.

Many changes took place at Al-Azhar University in the 1960's. Along with courses in Islamic religion, Islamic law, and Arabic studies, the school began teaching medicine, engineering, and agriculture. It also began to instruct women. In the 1970's, some Muslim fundamentalists revived a movement that believed the government had strayed from Islamic beliefs.

The great courtyard of Al-Azhar is surrounded by covered walkways, behind which lie the students' living quarters. The side of the courtyard facing the Arab city of Mecca contains the university's place of prayer, because Muslims face Mecca when they pray.

Students from many countries come to study Islam at Al-Azhar University, where learning and piety go hand in hand. All students are called *mugawireen,* which means *neighbors.*

Islam is a unifying bond in Egypt. To all Muslims, Islam is a total way of life, not just a religious creed. Muslims believe that the teachings of Islam must guide all of society, not just the individuals within the society.

However, different segments of Egypt's population have different opinions about their religion. The views of a scholarly ulama of Al-Azhar may be worlds apart from those of a poor rural peasant. The ulama follows the *orthodox,* or traditional, Islamic beliefs of the holy book of Islam. Less educated people, however, may also follow quite nontraditional practices, such as honoring saints or believing in the power of charms and the evil eye. The fundamentalists insist that all of Egyptian society should follow their view, and many fundamentalists would like to see the Islamic code of laws called *Shari'ah* become the law of all Egypt.

Because the same kind of religious conflict is taking place in other Muslim countries in North Africa and the Middle East today, Egypt—the traditional center of Islamic learning and culture—holds an increasingly important position in today's Muslim world. Now, more than 1,000 years after its inception, Al-Azhar University remains the most influential religious school in Islam.

# CAIRO

The Egyptian capital of Cairo is the largest city in Africa. More than 7.7 million people live within the city limits.

Cairo is a fascinating mixture of the old and the new. Many of the more modern sections of the city lie on the west bank of the Nile River, on islands in the river, and in Garden City—a narrow strip of land on the river's east bank. Bridges built during the 1900's to connect the riverbanks and islands have made modern urban development possible.

Today, the western part of Cairo includes many government buildings, foreign embassies, hotels, museums, and universities. These newer areas are studded with gardens, parks, and public squares.

In eastern Cairo, by contrast, stand the oldest and most historic sections of the city. In these old quarters, narrow, winding streets meander past centuries-old buildings. On some streets, in colorful outdoor markets called *bazaars,* almost every available space is crowded with goods.

The old sections are also known for their mosques, which include many outstanding examples of Muslim

A street scene in one of Cairo's old sections features open-front shops and pedestrians making way for donkey carts.

architecture. Some are hundreds of years old, including the mosque of Ahmed Ibn Tulun, built in the A.D. 870's.

At least one minaret can be seen from almost any spot in Cairo's old sections. From these slender towers atop the mosques, Muslim officials called *muezzins* summon the faithful to prayer five times every day.

The residents of Cairo are called Cairenes. Most Cairenes are Arab Muslims, but Christians called Copts also live in the city. The Copts trace their origin back to the Christians who lived in Egypt before the Arabs arrived in the 600's.

Well-to-do Cairenes and poor Cairenes lead vastly different lives. Most middle-class and wealthy people dress in clothes similar to those worn in North America and Europe. They live in Garden City, on the islands, or in the suburbs. These Cairenes are the professional people—doctors, lawyers, teachers, government officials, and managers.

In the old sections on the eastern side of Cairo, many people live in small, crowded apartments on the upper floors of run-down buildings. Numerous minarets rise above the rooftops.

# EGYPT'S METROPOLIS

Cairo sprawls over the banks of the Nile River, covering about 83 square miles (215 square kilometers), but its west and east sides seem worlds apart. Western Cairo is more modern, with many buildings that date from the 1900's. Eastern Cairo is much older and more historic.

Cairo experienced tremendous growth during the 1900's. Its population became more stable in the 1980's, but the population of the surrounding metropolitan area continued to increase.

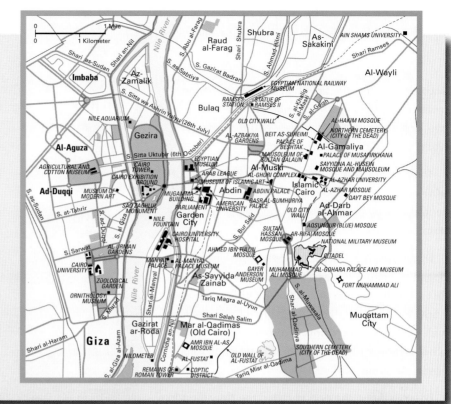

Industry and tourism flourish side by side in Cairo. Modern office towers in the newer sections of the city rise above the Nile, while a large ferryboat carries passengers upstream toward the monuments of ancient Egypt.

A young Egyptian girl weaves a rug in Sakkara, near Cairo. The Egyptian people have a rich artistic heritage, dating from the fine paintings and statues created in ancient times.

Poor Cairenes are crowded into small apartments or makeshift huts in the old sections. Some of the poorest people have taken refuge in old tombs in an area on the outskirts known as the City of the Dead. Many poor Cairenes do unskilled work in factories or small shops, and others are unemployed. Some dress in traditional Egyptian robes.

The population of Cairo has increased dramatically over the last 100 years. This rapid growth was caused by several factors, including Egypt's high birth rate, the movement of rural people to the city, and the influx of refugees who sought shelter in Cairo after their towns were damaged during Arab-Israeli wars.

The rapid growth of Cairo has intensified the city's problems. As Cairenes cope with housing shortages and traffic jams, the number of poor people continues to grow.

# HISTORY

According to legend, Egyptian civilization began more than 5,000 years ago when King Menes united two kingdoms on the Nile River and formed the world's first national government.

The age of great pyramid building began about 400 years later, in about 2650 B.C. This period of central rule by powerful kings, called the Old Kingdom, lasted for 500 years.

During the 1900's B.C., Middle Kingdom kings began to restore the wealth and power that Egypt lost after the decline of the Old Kingdom. But about 1630 B.C., Asian settlers on the Nile Delta—armed with horse-drawn chariots, improved bows, and other weapons—seized control. Their kings ruled Egypt for about 100 years.

Egypt entered the period called the New Kingdom about 1539 B.C. It became the world's strongest power over the next 500 years. Leaders of that time included Queen Hatshepsut, who led armies in battle, and King Thutmose III, who expanded the Egyptian empire into southwest Asia.

Ancient Egypt began to decline rapidly after about 1075 B.C., and in 332 B.C., Alexander the Great of Macedonia added Egypt to his empire. When he died in 323 B.C., Ptolemy, a general in his army, became king of Egypt.

Queen Cleopatra VII, a descendant of Ptolemy, lost Egypt to Rome when she married and joined forces with Mark Antony, a co-ruler of Rome. Antony and Cleopatra's navy was defeated by the fleet of Octavian, another Roman co-ruler, and Egypt became a Roman province.

Between A.D. 639 and 642, Arab Muslims began to conquer the Egyptians and convert them to Islam. Thus, Egypt became part of the Muslim Empire. It was run by rulers called caliphs.

By the mid-1100's, Christian crusaders from Europe were threatening Egypt as part of their effort to recapture the Holy Land from the Muslims. When the Egyptian caliph asked Syrian Muslims for help, Saladin, an officer in the Syrian army, drove the crusaders out and became sultan, or prince, of Egypt.

The Ahmed Ibn Tulun Mosque in Cairo, built in the A.D. 870's, stands as a reminder of Egypt's Islamic traditions.

In 1250, military slaves called Mamluks seized control of Egypt. Ruthless power struggles followed for more than 200 years. The Ottoman Turks invaded Egypt in 1517, but the Mamluks were able to keep much of their power until 1798. In that year, France's Napoleon Bonaparte defeated the Mamluks in the Battle of the Pyramids.

With aid from the British, the Ottomans forced the French out of Egypt in 1801. In 1805, in the disorder that followed the French departure, a Turkish officer named Muhammad Ali made himself ruler. He began an ambitious program to modernize Egypt, but many of his reforms failed.

| | |
|---|---|
| 3100 B.C. | The legendary King Menes unifies Egypt. |
| 2650–2150 B.C. | Old Kingdom—noted for pyramid building—flourishes. |
| 1975–1640 B.C. | During the Middle Kingdom, trade and arts flourish. |
| 1539–1075 B.C. | Empire developed during the New Kingdom. |
| 332 B.C. | Alexander the Great conquers Egypt. |
| 305 B.C. | Ptolemy founds dynasty. |
| 30 B.C. | Egypt becomes Roman province. |
| 642 | Arab Muslims conquer Egypt. |
| 969–1171 | Fatimid dynasty rules Egypt. |
| 1171–1250 | Ayyubid dynasty rules. |
| 1250–1517 | Mamluks rule. |
| 1517 | Ottoman Turks conquer Egypt. |
| 1798 | Napoleon conquers Egypt. |
| 1801 | France withdraws from Egypt. |
| 1805 | Muhammad Ali seizes control. |
| 1869 | Suez Canal opens. |
| 1875 | Egypt sells canal shares to the United Kingdom. |
| 1882 | British troops occupy Egypt. |
| 1914 | The United Kingdom declares Egypt a protectorate. |
| 1922 | The United Kingdom grants Egypt independence. |
| 1948-1949 | Egypt joins in Arab-Israeli war. |
| 1952 | King Faruk forced to step down. |
| 1953 | Egypt becomes a republic. |
| 1954 | Gamal Abdel Nasser comes to power. British agree to withdraw all troops from Egypt by June 1956. |
| 1956 | Suez Canal crisis. Israeli forces invade. British and French troops land, but are withdrawn by the end of the year. |
| 1958 | Egypt joins Syria in United Arab Republic. Syria withdraws in 1961. |
| 1967 | Israel defeats Egypt in Six-Day War. |
| 1968 | Aswan High Dam begins operation. |
| 1970 | Nasser dies. Anwar el-Sadat becomes president. |
| 1973 | Egyptian troops invade Israeli-held Sinai Peninsula in fourth Arab-Israeli war. |
| 1978 | Camp David Accords with Israel. |
| 1979 | Egypt and Israel sign peace treaty. |
| 1981 | Sadat assassinated. Hosni Mubarak becomes president. |
| 1991 | Egyptian troops take part in efforts to liberate Kuwait following Iraqi invasion. |
| 2007 | Voters approve amendments to the Constitution. |
| 2011 | Government is overthrown in a revolution. |

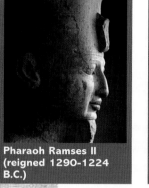

**Pharaoh Ramses II (reigned 1290-1224 B.C.)**

**Muhammad Ali (1769-1849)**

**Anwar Sadat (1918-1981)**

Under Muhammad Ali's son, Said, French interests began to cut a great canal through the Isthmus of Suez. Said's nephew, Ismail, sold Egypt's interests in the canal to the United Kingdom in 1875. British interests in Egypt increased, and in 1914 the United Kingdom made Egypt a protectorate and kept troops there—even after Egypt was given independence in 1922.

In 1952, Gamal Abdel Nasser and other army officers led an overthrow of Egypt's King Faruk. Under Nasser, Egypt took control of the Suez Canal and began a period of great economic and social change. The nation fought two wars with Israel and suffered a major defeat in 1967.

Anwar el-Sadat became Egypt's president when Nasser died. Sadat signed a treaty with Israel in 1979, but enraged Muslim fundamentalists killed Sadat in 1981. He was succeeded by Hosni Mubarak. Mubarak was overthrown in a revolution in 2011, and a military government took control of Egypt. Egypt held its first free presidential elections in 2012.

The temple of Abu Simbel was carved out of a mountainside beside the Nile River in southern Egypt in the 1200's B.C. Four seated figures of Pharaoh Ramses II guard the entrance. The temple was actually cut apart and moved to higher ground in the mid-1960's to avoid it being covered by the lake created by the Aswan High Dam.

# EL SALVADOR

Indigenous (American Indian) people have lived in what is now El Salvador for thousands of years. The ruins of huge limestone pyramids built by the Maya between A.D. 100 and 1000 still stand in western El Salvador. By the A.D. 900's, Pipil Indians began to migrate into the region from what is now Mexico. After the year 1000, the Pipil seized control of the lands west of the Lempa river. The Pipil built cities, raised crops, and produced skilled weaving.

## Spanish rule

The first Europeans in the region were Spanish soldiers led by Pedro de Alvarado, who invaded the land in 1524. After several years of fierce struggle, the Spanish conquered the Pipil.

El Salvador was a Spanish colony for the next 300 years, but the land had little gold or silver, so it attracted few colonists. Those who did come to the country mainly farmed the land and raised cattle.

With four other Central American states, El Salvador broke away from Spanish rule in 1821. The five states formed the United Provinces of Central America in 1823. José Matías Delgado, a Salvadoran Catholic priest, led El Salvador's revolt against Spain and drafted the constitution of the five-state union. However, the union began to break up in 1838, and El Salvador withdrew in 1840. In 1841, the country officially declared its independence.

## Struggle for reform

Political violence troubled El Salvador for the rest of the 1800's. Five of the country's presidents were overthrown in revolts, and two others were executed. Dictators from other Central American countries controlled the weak Salvadoran presidents.

In the late 1800's, the political situation became more stable. The country's economy also improved, as coffee and other farm products were grown and exported. Still, great social inequality persisted. Most of the nation's best farmland belonged to a small

In El Salvador, many Indian children and their families live in poverty. Much of the population lacks access to basic health services, and urban slums continue to grow.

Government soldiers search for guerrillas in a poor section of San Salvador. In November 1989, more than 1,300 people died in San Salvador during large-scale fighting between rebels and government troops. The war ended in 1992.

group of wealthy people, while many rural people had poor farmland—or no land at all.

In 1931, General Maximiliano Hernández Martínez seized control of the government. During his rule, he built public schools, expanded social services, and supported labor reform. However, he ruled as a dictator, and a revolution led by landowners, members of the middle class, professionals, and university students deposed him in 1944.

In 1956, another army officer, Colonel José María Lemus, won the presidency. However, some people disputed the election results. When Salvadorans began to demand election reform, Lemus banned free speech and imprisoned his opponents. A military group overthrew him in 1960. A *junta* (council) then ruled the country until 1961, when it was ousted by another junta.

In 1962, the Salvadoran legislature adopted a new constitution, and the people elected Colonel Julio Adalberto Rivera as president. Rivera's government raised the income taxes of the wealthy and gave some public and private lands to poor families, but problems continued.

Widespread protests broke out in the late 1970's, with demands for land and jobs for the poor. Some Roman Catholic priests and nuns supported these demands.

In 1979, army officers overthrew the president and replaced him with a junta. Widespread rioting took place when Archbishop Óscar Arnulfo Romero was assassinated. Romero was a champion of El Salvador's poor and had openly criticized the government.

With the violence increasing, the junta appointed José Napoleón Duarte as president in 1980. Duarte began reforms, but protests continued. By the early 1980's, a civil war between government troops and rebel guerrilla forces was underway. The United States, under President Ronald Reagan, increased its financial and military aid to the government of El Salvador.

El Salvador adopted a new constitution in 1983. Voters elected Duarte president in the 1984 election.

In January 1992, the government and guerrilla leaders signed a peace treaty providing for an end to the 12-year civil war. The war officially ended on Dec. 15, 1992.

In 2001, earthquakes in El Salvador killed about 1,200 people and left more than 1 million without homes. In 2004, El Salvador ratified the Dominican Republic-Central America-United States Free Trade Agreement (CAFTA-DR), which aimed to reduce trade barriers between the participating countries.

Rescue workers search for victims of Hurricane Ida in 2009 in the city of San Vicente. The hurricane triggered floods and mudslides that killed more than 120 people.

# EL SALVADOR TODAY

El Salvador is a tiny, but crowded, Central American country. The smallest nation in Central America in area, El Salvador has more people than any other country in the region except Guatemala and Honduras. El Salvador ranks as the most densely populated country on the mainland of the Americas.

## The land and economy

Most of the country's fertile farmland lies in the Central Region. This region is a broad plateau of gently rolling land. It is bordered on the south by the Coastal Range, a rugged mountain range with high, inactive volcanoes. On the range's lower slopes, coffee plantations and cattle ranches sprawl among forests of oak and pine trees. Coffee is one of El Salvador's leading crops.

Past eruptions of the Coastal Range volcanoes helped create the fertile soil of the Central Region plateau, now El Salvador's chief agricultural region. Many Salvadoran farmers own small farms and grow beans, corn, rice, and other crops to feed their families or to trade or sell at local markets. Other farmers work on large commercial plantations called *fincas*, which raise coffee and sugar cane.

The Central Region is home to most of the people of El Salvador. About 75 percent of the Salvadoran people live there, especially in the large cities, such as San Salvador and Santa Ana. Many rural people have moved to the cities looking for work because the supply of farmland has become so limited.

Most of the country's industry is in this area. The federal government has encouraged new industries in an attempt to reduce the country's dependence on agriculture. Manufacturing now composes one of the leading sectors in El Salvador's economy. The leading industries produce chemicals, clothing, foods and beverages, medicine, petroleum products, and textiles.

South of the Coastal Range, the Coastal Lowlands, a narrow strip of fertile land, stretches along the Pacific shore. Grasslands, swamps, and tropical forests

## FACTS

| | |
|---|---|
| Official name: | República de El Salvador (Republic of El Salvador) |
| Capital: | San Salvador |
| Terrain: | Mostly mountains with narrow coastal belt and central plateau |
| Area: | 8,124 mi² (21,041 km²) |
| Climate: | Tropical; rainy season (May to October); dry season (November to April); tropical on coast; temperate in uplands |
| Main rivers: | Lempa, Río Grande de San Miguel |
| Highest elevation: | Cerro El Pital, 8,957 ft (2,730 m) |
| Lowest elevation: | Pacific Ocean, sea level |
| Form of government: | Republic |
| Head of state: | President |
| Head of government: | President |
| Administrative areas: | 14 departamentos (departments) |
| Legislature: | Asamblea Legislativa (Legislative Assembly) with 84 members serving three-year terms |
| Court system: | Corte Suprema (Supreme Court) |
| Armed forces: | 15,500 troops |
| National holiday: | Independence Day - September 15 (1821) |
| Estimated 2010 population: | 7,191,000 |
| Population density: | 885 persons per mi² (342 per km²) |
| Population distribution: | 60% urban, 40% rural |
| Life expectancy in years: | Male, 68; female, 75 |
| Doctors per 1,000 people: | 1.2 |
| Birth rate per 1,000: | 24 |
| Death rate per 1,000: | 6 |
| Infant mortality: | 22 deaths per 1,000 live births |
| Age structure: | 0-14: 34%; 15-64: 61%; 65 and over: 5% |
| Internet users per 100 people: | 12 |
| Internet code: | .sv |
| Languages spoken: | Spanish, Nahua |
| Religions: | Roman Catholic 57.1%, Protestant 21.2%, Jehovah's Witness 1.9%, Mormon 0.7%, other 19.1% |
| Currency: | Salvadoran colon and United States dollar |
| Gross domestic product (GDP) in 2008: | $22.17 billion U.S. |
| Real annual growth rate (2008): | 3.2% |
| GDP per capita (2008): | $3,071 U.S. |
| Goods exported: | Coffee, ethanol, medicine, sugar, textiles |
| Goods imported: | Chemicals, electronics, food, machinery, motor vehicles, petroleum |
| Trading partners: | Costa Rica, Guatemala, Honduras, Mexico, Nicaragua, United States |

El Salvador is the smallest Central American country—about the same size as Massachusetts. Guatemala lies to the northwest; Honduras lies east and northeast.

El Salvador's Pacific coast is only 189 miles (304 kilometers) long. Fishing crews catch shrimp and lobsters in the offshore waters.

once covered much of this plain, but large sections have been developed for farmland. Cotton and sugar cane thrive in these warm, humid lowlands. A number of factories and a fishing industry are located near the town of Acajutla.

The Interior Highlands to the north of the Central Region are made up mainly of a low mountain range called the Sierra Madre. Hardened lava, rocks, and volcanic ash cover much of this area. The Lempa, El Salvador's longest river, flows 200 miles (320 kilometers) from the Sierra Madre to the Pacific. The highlands are El Salvador's most thinly populated region, with only a few small farms and ranches.

## Civil war

A civil war between the government and the leftist Farabundo Martí National Liberation Front (FMLN) claimed about 75,000 lives between 1979 and 1992. Many atrocities were committed during the war, especially by the military. In January 1992, the government and FMLN signed a peace agreement.

A United Nations Truth Commission report found that most serious acts of violence committed during the war had been carried out by Salvadoran troops, paramilitary forces, and right-wing death squads. The Salvadoran governments of the 1990's argued that the report had exaggerated the amount of state violence against civilians.

# PEOPLE

More than 7.1 million people live in El Salvador. More than 90 percent of all Salvadorans are *mestizos,* or people of mixed European and indigenous descent. About 7 percent identify themselves as white, and less than 3 percent are indigenous.

## Two cultures

The people of El Salvador form two distinct cultural groups—Ladinos and Indians. Ladinos follow Spanish-American ways of life and speak Spanish. All mestizos and whites are Ladinos, but Indians can also become Ladinos if they adopt the Spanish-American culture, as have many Indians.

Many other Indians still follow the ways of their ancestors. El Salvador's Pancho Indians, who are descended from the Pipil Indians, live in villages built by the Pipil and speak a form of Nahua, the Pipil language. The Pipil were the largest Indian group in El Salvador when the Spaniards first arrived. Most Pancho Indians now live in the southwest.

Faces in the crowd at an election rally in San Salvador show the Salvadoran people's mixed Indian and Spanish ancestry.

A guerrilla fighter guards a rebel base during the Salvadoran civil war of the 1980's and 1990's. In a country where most of the farmland traditionally has been controlled by a few wealthy landowners, the rebels sought greater equality in land ownership.

## Social class and family life

About 40 percent of all Salvadorans live on farms or in rural areas. Some farmers live in adobe houses with dirt floors and thatched roofs. Many poor people in rural areas live in *wattle houses.* The walls of such houses are made of tree branches woven together and covered with mud. In sharp contrast, some wealthy owners of coffee plantations live in luxury on their scenic estates.

Since the 1940's, hundreds of thousands of rural Salvadorans have moved to the cities to look for jobs, but there is not enough work for them all. Entire families often live in one-room apartments in crowded, run-down buildings. Middle-class city residents live in row houses or in comfortable apartments. The rich live in the suburbs of the cities in luxurious, modern houses with landscaped gardens.

However, many of these people are now moving to the cities.

In El Salvador, widespread poverty has disrupted family life for many people. Some men move from place to place seeking work, and many of El Salvador's children live in orphanages or wander the streets or countryside with no one to care for them.

Public education in El Salvador is inadequate. Children in poor rural areas have less opportunity to attend school than in urban areas. Students who complete nine years of elementary school can go on to secondary school for three years and then attend a university. However, most universities are concentrated in San Salvador, and few rural Salvadorans are able to attend.

## Leisure

The people of El Salvador love to spend their leisure time outdoors. Soccer is the national sport, and a game is often in progress in neighborhood fields. Families like to spend their weekends at resorts near lakes or on Pacific beaches. Los Chorros, a popular national park near Nueva San Salvador, consists of a series of natural pools surrounded by tropical gardens and waterfalls.

Almost 60 percent of all Salvadorans are Roman Catholic, and religious festivals are popular. The most colorful festival of the year celebrates the Feast of the Holy Savior of the World. From July 24 to August 6, the people celebrate this festival with carnival rides, fireworks, folk dancing, and colorful processions.

## Food

The diet of most Salvadorans consists mainly of beans, bread, corn, and rice. Dairy products and meat are occasional luxuries when the people can afford to buy them.

Salvadorans have bread and coffee for breakfast and eat their main meal at midday. In the late afternoon, many people snack on *pupusas,* which are corn meal cakes stuffed with chopped meat, beans, and spices.

An Easter procession reflects the importance of religious festivals in El Salvador.

Fishing boats are hoisted out of the water at the end of the day at the town of La Libertad, one of El Salvador's Pacific ports. The catch includes shrimp and lobster, but seafood is not a major item in the diet of most Salvadorans.

# EQUATORIAL GUINEA

The small western African country of Equatorial Guinea consists of a mainland region and five islands in the Atlantic Ocean. The mainland territory, called Río Muni, is home to most of the nation's people. The largest island, Bioko, lies about 100 miles (160 kilometers) northwest of Río Muni, while the other four islands—Corisco, Elobey Chico, Elobey Grande, and Annobón—lie southwest of Río Muni.

## People

Equatorial Guinea has a little more than half a million people, and about 60 percent live in rural areas. Most rural people are farmers, raising such food crops as bananas, cassava, and sweet potatoes. Some farms produce cacao for export, especially on Bioko. Coffee also thrives in Bioko's rich volcanic soil.

Some rural people fish for a living, especially on the islands, and others work in lumber camps. Dense tropical rain forests cover much of the land, and forestry is important to the country's economy. The okoumé tree, for example, is cut down to make plywood.

About 40 percent of the people of Equatorial Guinea live in urban areas, such as the largest city, Bata, and the capital, Malabo. Many of these people work in small industries or in trade. The country has some food-processing plants, but little manufacturing. Petroleum production is the chief economic activity in Equatorial Guinea and accounts for more than 90 percent of the country's exports. Large oil and gas deposits were discovered near Bioko in the mid-1900's.

The people of Equatorial Guinea belong mainly to various African ethnic groups. In Río Muni, most people are members of the Fang group. Their language—also called Fang—is the most widely spoken language in the country, though Spanish and French are the official languages. Most of the people on the island of Bioko belong to the Fernandino or Bubi ethnic groups.

Equatorial Guinea suffers from poor educational and medical services. Many children do not attend school because of a shortage of teachers, and diseases such as malaria and measles spread quickly due to the lack of medical care.

## FACTS

| | |
|---|---|
| Official name: | Republica de Guinea Ecuatorial (Republic of Equatorial Guinea) |
| Capital: | Malabo |
| Terrain: | Coastal plains rise to interior hills; islands are volcanic |
| Area: | 10,831 mi² (28,051 km²) |
| Climate: | Tropical; always hot, humid |
| Main river: | Mbini |
| Highest elevation: | Pico Basile, 9,869 ft (3,008 m) |
| Lowest elevation: | Sea level along the coasts |
| Form of government: | Republic |
| Head of state: | President |
| Head of government: | Prime minister |
| Administrative areas: | 7 provincias (provinces) |
| Legislature: | Camara de Representantes del Pueblo (House of People's Representatives) with 100 members serving five-year terms |
| Court system: | Supreme Tribunal |
| Armed forces: | 1,300 troops |
| National holiday: | Independence Day - October 12 (1968) |
| Estimated 2010 population: | 598,000 |
| Population density: | 55 persons per mi² (21 per km²) |
| Population distribution: | 60% rural, 40% urban |
| Life expectancy in years: | Male, 60; female, 61 |
| Doctors per 1,000 people: | 0.3 |
| Birth rate per 1,000: | 38 |
| Death rate per 1,000: | 10 |
| Infant mortality: | 86 deaths per 1,000 live births |
| Age structure: | 0-14: 42%; 15-64: 54%; 65 and over: 4% |
| Internet users per 100 people: | 1.7 |
| Internet code: | .gq |
| Languages spoken: | Spanish (official), French (official), Pidgin English, Fang, Bubi, Ibo |
| Religions: | Roman Catholic (predominant), Protestant |
| Currency: | Coopération Financière en Afrique Centrale franc |
| Gross domestic product (GDP) in 2008: | $18.53 billion U.S. |
| Real annual growth rate (2008): | 11.2% |
| GDP per capita (2008): | $34,443 U.S. |
| Goods exported: | Mostly: petroleum Also: chemicals, cocoa, timber |
| Goods imported: | Machinery, petroleum products, transportation equipment |
| Trading partners: | China, Spain, Taiwan, United States |

## History

Pygmies, who inhabited the area before the 1200's, were probably the first people to live in what is now mainland Río Muni. Then other groups began to move into the region, including the Fang, the Benga, and the Bubi. Bubi from the mainland also settled on Bioko in the 1200's.

The first Europeans in the region were the Portuguese, who landed on the island of Annobón in 1471. Portugal later claimed that island, along with Bioko and part of the mainland coast. Spain gained control of these territories in the mid-1800's, and the region became a Spanish colony in 1959.

Equatorial Guinea became an independent nation on Oct. 12, 1968. Later that year, Francisco Macías Nguema took control as president and dictator. In 1979, army officers overthrew Macías and established a military government. Macías was executed, and Lieutenant Colonel Teodoro Obiang Nguema Mbasogo became president.

Government by the military continued until 1989, when Obiang Nguema was elected president in an election in which his Democratic Party for Equatorial Guinea was the only political party allowed. In 1991, a new constitution was approved. In multiparty elections held in 1996, 2002, and 2009, Obiang Nguema was reelected president. However, the elections were widely viewed as fraudulent.

A treelined riverbank is silhouetted by light from the setting sun. Dense tropical rain forests cover much of the country.

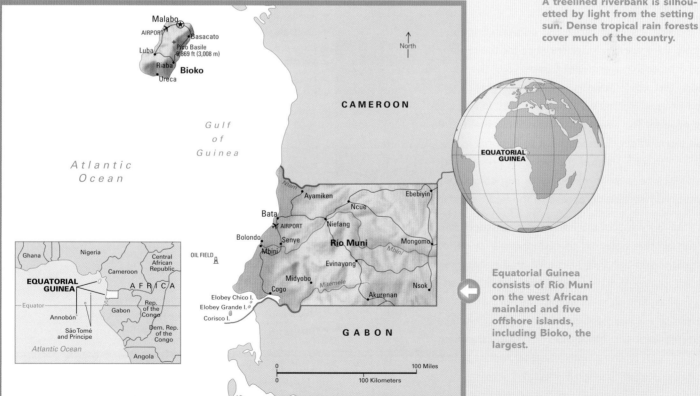

Equatorial Guinea consists of Río Muni on the west African mainland and five offshore islands, including Bioko, the largest.

# ERITREA

Eritrea is a small country situated on the east coast of Africa. It is bordered to the north and west by Sudan, to the east by the Red Sea, and to the south by Ethiopia and Djibouti. Formerly a province of Ethiopia, Eritrea gained its independence in 1993 after a 30-year struggle.

Eritrea's coastal plain is one of the hottest and driest spots in Africa. Although only a small portion of the land is under cultivation, about 80 percent of the people work in agriculture. The nation's chief products include barley and lentils.

Eritrea was colonized by Italy in the late 1800's. The Italians used the region as a base from which they established their brief rule over Ethiopia. They built railways and other commercial enterprises in Eritrea. This period of colonization helped the Eritreans develop many skills—and an important trading edge over Ethiopia.

The United Kingdom drove Italy from Eritrea in 1941, during World War II. In 1952, British administration of the territory ended, and Eritrea was joined with Ethiopia as part of a federation. But the region governed itself.

In 1961, civil war broke out between Eritrean rebels, who wanted independence for the region, and Ethiopian government troops. Over the next 30 years, the rebels developed into the highly organized Eritrean People's Liberation Front, led by Issaias Afwerki. The separatists maintained that Eritrea had been unfairly handed back to Ethiopia in 1952. Ethiopia resisted the move for Eritrean independence because it feared becoming completely landlocked and losing access to the Red Sea.

After 30 years of violence and bloodshed, Eritrean fighters defeated the Ethiopian army—the largest in sub-Saharan Africa. When the struggle ended in May 1991—shortly after the Ethiopian army's last strongholds in the cities of Asmara and Keren fell—it had become modern Africa's longest and deadliest war. More than 100,000 Ethiopian and Eritrean fighters were killed, as well as tens of thousands of civilians in Eritrea.

## FACTS

| | |
|---|---|
| Official name: | Hagere Ertra (State of Eritrea) |
| Capital: | Asmara |
| Terrain: | Dominated by extension of Ethiopian north-south trending highlands, descending on the east to a coastal desert plain, on the northwest to hilly terrain, and on the southwest to flat-to-rolling plains |
| Area: | 45,406 mi² (117,600 km²) |
| Climate: | Hot, dry desert strip along Red Sea coast; cooler and wetter in the central highlands; semiarid in western hills and lowlands; rainfall heaviest June to September except in coastal desert |
| Main rivers: | N/A |
| Highest elevation: | Mount Soira, 9,885 ft (3,013 m) |
| Lowest elevation: | Near Kulul within the Denakil Depression, 246 ft (75 m) below sea level |
| Form of government: | Transitional government |
| Head of state: | President |
| Head of government: | President |
| Administrative areas: | 6 regions |
| Legislature: | National Assembly with 150 members serving five-year terms |
| Court system: | Supreme Court, provincial courts, district courts |
| Armed forces: | 201,800 troops |
| National holiday: | Independence Day - May 24 (1993) |
| Estimated 2010 population: | 5,338,000 |
| Population density: | 118 persons per mi² (45 per km²) |
| Population distribution: | 79% rural, 21% urban |
| Life expectancy in years: | Male, 57; female, 61 |
| Doctors per 1,000 people: | Less than 0.05 |
| Birth rate per 1,000: | 39 |
| Death rate per 1,000: | 9 |
| Infant mortality: | 46 deaths per 1,000 live births |
| Age structure: | 0-14: 43%; 15-64: 55%; 65 and over: 2% |
| Internet users per 100 people: | 3 |
| Internet code: | .er |
| Languages spoken: | Tigrinya (official), Afar, Arabic, Tigre and Kunama, other Cushitic languages |
| Religions: | Christian 50%, Muslim 48%, indigenous beliefs 2% |
| Currency: | Nakfa |
| Gross domestic product (GDP) in 2008: | $1.48 billion U.S. |
| Real annual growth rate (2008): | 2.0% |
| GDP per capita (2008): | $303 U.S. |
| Goods exported: | Food, leather, small manufactures |
| Goods imported: | Food, machinery, petroleum products |
| Trading partners: | China, Italy, Saudi Arabia |

After the war ended, the Eritreans began the long and difficult task of rebuilding a country devastated by three decades of fighting. Schools, roads, and medical facilities were rebuilt. Elections at the village, local, and provincial level were held, and work on a constitution began. Eritrean leaders also announced that their country would have a market-driven economy.

After an April 1993 referendum, in which the Eritreans overwhelmingly voted for independence, the nation declared its independence on May 24, 1993. A four-year transitional period, during which the country moved toward a constitutional political system, was also declared. In 1997, Eritrea adopted a new constitution.

A section of the border between Eritrea and Ethiopia was not clearly defined when Eritrea achieved independence. In May 1998, the two nations began warring over this disputed area. The two sides signed a cease-fire agreement in 2000, and they later signed a formal peace treaty. A commission identified the border between the countries in 2002.

In August 2009, an international court based in The Hague, in the Netherlands, blamed Eritrea for starting the border war and awarded damages to both sides. Both countries accepted the ruling, which left Eritrea owing Ethiopia $10 million.

Droughts struck Eritrea in the early 2000's. As a result, more than 2 million people faced starvation.

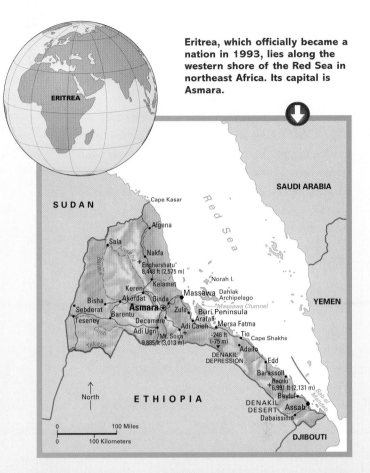

Eritrea, which officially became a nation in 1993, lies along the western shore of the Red Sea in northeast Africa. Its capital is Asmara.

Jubilant Eritreans celebrate the passage of an April 1993 referendum that established Eritrea as an independent nation.

The valley of the Anseba River is located in an inland region of Eritrea, in the west of the country. The climate in this area is semiarid.

# ESTONIA

Estonia is a European nation that gained its independence from the Soviet Union in 1991. Estonia's landscape consists chiefly of a low plain covered by farmland, forests, and swamps. The country has 481 miles (774 kilometers) of coastline along the Baltic Sea, the Gulf of Finland, and the Gulf of Riga. Estonia also includes several Baltic islands, the largest of which is Saaremaa Island.

When the Soviet Union took over Estonia in 1940, the Estonians—a people related to the Finns—made up about 90 percent of the population. However, many Russians emigrated to the republic to fill jobs created by rapid industrialization following World War II (1939–1945). As a result, native Estonians now account for only about 70 percent of the population.

## History

The ancestors of present-day native Estonians settled in the area several thousand years ago. By the 1500's, German nobles owned much of Estonia. Sweden took over northern Estonia in 1561, and Poland conquered the southern part of the country.

Sweden controlled all of Estonia between 1625 and 1721, when Russia took over the area. In the mid-1800's, a movement for Estonian independence gained strength. In 1918, Estonia proclaimed its independence and established a democratic government.

## From independence to annexation

In 1939, Soviet dictator Joseph Stalin and Adolf Hitler, the leader of Nazi Germany, secretly agreed to divide up between themselves a number of Eastern European countries. In August 1940, the Soviet Union forcibly annexed Estonia and turned the nation's private factories and farms into government-controlled enterprises.

## FACTS

| | |
|---|---|
| **Official name:** | **Eesti Vabariik (Republic of Estonia)** |
| **Capital:** | **Tallinn** |
| **Terrain:** | **Marshy, lowlands** |
| **Area:** | **17,463 mi² (45,228 km²)** |
| **Climate:** | **Maritime, wet, moderate winters, cool summers** |
| **Main rivers:** | **Gauja, Jägala, Kasari, Narva** |
| **Highest elevation:** | **Munamägi, 1,043 ft (318 m)** |
| **Lowest elevation:** | **Baltic Sea, sea level** |
| **Form of government:** | **Parliamentary republic** |
| **Head of state:** | **President** |
| **Head of government:** | **Prime minister** |
| **Administrative areas:** | **15 maakonnad (counties)** |
| **Legislature:** | **Riigikogu (Parliament) with 101 members serving four-year terms** |
| **Court system:** | **National Court** |
| **Armed forces:** | **5,300 troops** |
| **National holiday:** | **Independence Day - February 24 (1918)** |
| **Estimated 2010 population:** | **1,321,000** |
| **Population density:** | **76 persons per mi² (29 per km²)** |
| **Population distribution:** | **69% urban, 31% rural** |
| **Life expectancy in years:** | **Male, 67; female, 78** |
| **Doctors per 1,000 people:** | **3.3** |
| **Birth rate per 1,000:** | **12** |
| **Death rate per 1,000:** | **13** |
| **Infant mortality:** | **5 deaths per 1,000 live births** |
| **Age structure:** | **0-14: 15%; 15-64: 68%; 65 and over: 17%** |
| **Internet users per 100 people:** | **64** |
| **Internet code:** | **.ee** |
| **Languages spoken:** | **Estonian (official), Russian, Ukrainian, Belarusian, Finnish** |
| **Religions:** | **Evangelical Lutheran 13.6%, Orthodox 12.8%, other Christian 1.4%, other 72.2%** |
| **Currency:** | **Estonian kroon** |
| **Gross domestic product (GDP) in 2008:** | **$23.26 billion U.S.** |
| **Real annual growth rate (2008):** | **-3.0%** |
| **GDP per capita (2008):** | **$17,437 U.S.** |
| **Goods exported:** | **Electronics, machinery, petroleum products, transportation equipment, wood products** |
| **Goods imported:** | **Electronics, machinery, petroleum products, textiles, transportation equipment** |
| **Trading partners:** | **Finland, Germany, Latvia, Lithuania, Russia, Sweden** |

The Estonians bitterly opposed the takeover. However, tens of thousands of people who joined the Estonian resistance movement were deported to Siberia. Resistance to Soviet control grew during the late 1980's, and Estonia began to press the Soviet Union for restoration of the republic's independence. In January 1991, the Soviet government responded to Estonia's suspension of compulsory military service by sending in troops to enforce the draft.

## Freedom reclaimed

After becoming independent in 1991, Estonia moved forward with economic reform and reduced government control of most economic activities. By the mid-1990's, most businesses had become privately owned. Estonia sought to strengthen its ties with Western Europe and to reduce Russian influence over its affairs. In 2004, Estonia joined both the European Union (EU) and the North Atlantic Treaty Organization (NATO). In 2011, Estonia joined the eurozone, the group of EU nations that use the euro as a common currency.

**The spires and turrets rising above Tallinn are typical of the impressive architecture for which the city is renowned.**

**Estonia lies on the Baltic Sea in northern Europe. It is one of the three Baltic States, along with Lithuania and Latvia.**

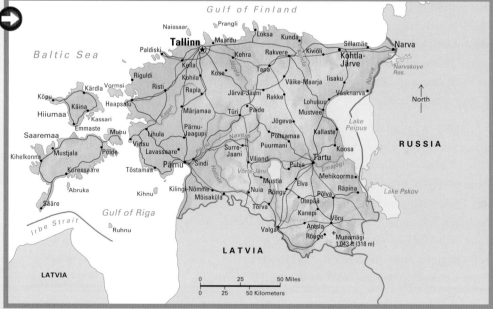

# ETHIOPIA

Ethiopia is one of the oldest nations in Africa. By the A.D. 200's, a powerful kingdom had been established in Ethiopia. Its capital, Aksum, grew wealthy by trading gold, ivory, and spices with Arabia, Egypt, Greece, India, Persia, and Rome.

The power of Aksum fell in the 600's, when Muslims took control of Arabia, the Red Sea, and the coast of northern Africa. In 1137, the Zagwé dynasty rose to power but was overthrown in 1270. After the 1500's, the Ethiopian Empire broke up into a number of small kingdoms.

The empire was rebuilt in the late 1800's, when Menelik II became emperor. In 1916, Menelik's daughter, Zauditu, became empress, ruling with the help of Ras Tafari, the son of Menelik's cousin. When Zauditu died in 1930, Tafari became emperor, taking the title Haile Selassie I.

In 1974, army officers overthrew Haile Selassie. Under Lieutenant Colonel Mengistu Haile-Mariam, the government adopted socialism and began land reform. The military leaders also began killing many of their opponents.

In the late 1970's, rebellion broke out in the northern region of Tigray, and Ethiopia went to war over Ogaden with its neighbor Somalia, which also claimed the region. The war with Somalia ended in 1988. Meanwhile, Eritrean rebels set up their own government in the Eritrean province and in 1993 gained independence from Ethiopia. From 1998 to 2000, Eritrea and Ethiopia battled over the location of their shared border.

In 1991, a group of rebel armies overthrew Mengistu's military government. The rebels formed a transitional government until elections could take place. In 1994, Ethiopia adopted a new constitution. Free elections took place in May 1995, and the ruling Ethiopian People's Revolutionary Democratic Front (EPRDF) won a majority. The EPRDF won elections again in 2005 and 2010. In both cases, opposition groups claimed the results were fraudulent.

## FACTS

| | |
|---|---|
| Official name: | Ityop'iya Federalawi Demokrasiyawi Ripeblik (Federal Democratic Republic of Ethiopia) |
| Capital: | Addis Ababa |
| Terrain: | High plateau with central mountain range divided by Great Rift Valley |
| Area: | 435,186 mi² (1,127,127 km²) |
| Climate: | Tropical monsoon with wide topographic-induced variation |
| Main rivers: | Awash, Baro, Blue Nile (Abay), Genale, Omo, Wabe Shebele |
| Highest elevation: | Ras Dashen, 15,158 ft (4,620 m) |
| Lowest elevation: | Denakil Depression, 381 ft (116 m) below sea level |
| Form of government: | Republic |
| Head of state: | President |
| Head of government: | Prime minister |
| Administrative areas: | 9 ethnically based stedader akababiwach (administrative regions), 2 chartered cities |
| Legislature: | Parliament consisting of the Council of the Federation or upper chamber with 118 members serving five-year terms and the Council of People's Representatives or lower chamber with 548 members serving five-year terms |
| Court system: | Supreme Court |
| Armed forces: | 138,000 troops |
| National holiday: | National Day - May 26 (1991) |
| Estimated 2010 population: | 88,013,000 |
| Population density: | 202 persons per mi² (78 per km²) |
| Population distribution: | 84% rural, 16% urban |
| Life expectancy in years: | Male, 50; female, 54 |
| Doctors per 1,000 people: | Less than 0.05 |
| Birth rate per 1,000: | 40 |
| Death rate per 1,000: | 13 |
| Infant mortality: | 77 deaths per 1,000 live births |
| Age structure: | 0-14: 44%; 15-64: 53%; 65 and over: 3% |
| Internet users per 100 people: | 0.4 |
| Internet code: | .et |
| Languages spoken: | Amharic (official), Tigrinya, Arabic, Guaragigna, Oromifa, English, Somali, other local languages |
| Religions: | Sunni Muslim, Ethiopian Orthodox Christian, Protestant, indigenous beliefs |
| Currency: | Birr |
| Gross domestic product (GDP) in 2008: | $25.74 billion U.S. |
| Real annual growth rate (2008): | 8.5% |
| GDP per capita (2008): | $329 U.S. |
| Goods exported: | Coffee, gold, leather products, oil seeds, vegetables |
| Goods imported: | Food, machinery, motor vehicles, petroleum products |
| Trading partners: | China, Germany, India, Italy, Saudi Arabia |

In December 2006, Ethiopian forces assisted Somali government forces in fighting back an Islamic militia that had gained control of southern Somalia. In January 2009, Ethiopian troops pulled out of Somalia, though fighting continued.

Although much of Ethiopia is fertile, farming methods are inefficient, and farmers often struggle to provide enough food for their families. Many farmers work the land with wooden plows pulled by oxen. In the 1970's, the 1980's, and the early 2000's, the country experienced several severe droughts, which led to the deaths of many people.

Nevertheless, agriculture is the nation's chief economic activity and employs most Ethiopian workers. Farmers' main food crops for their own use include corn, sorghum, *teff* (a grain), and wheat. Most farmers also raise cattle, chickens, goats, and sheep. Cash crops include coffee, khat, oilseeds, and sugar cane.

Service industries—especially banking, government, insurance, retail trade, tourism, and transportation—play a major role in Ethiopia's economy. Manufacturing forms only a small part of the economy.

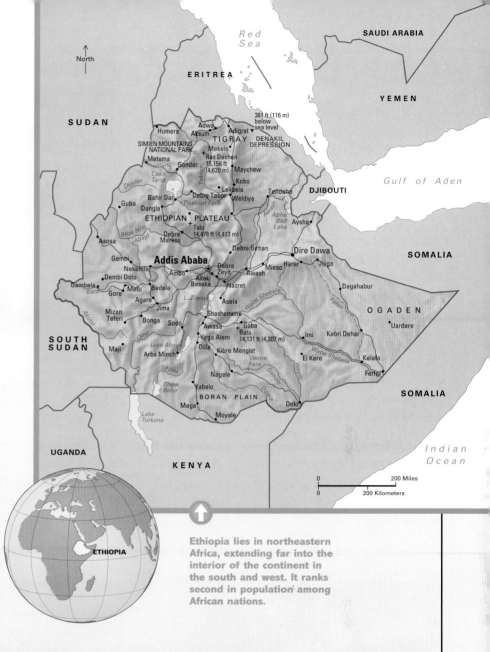

Ethiopia lies in northeastern Africa, extending far into the interior of the continent in the south and west. It ranks second in population among African nations.

Rolling green hills in northwestern Ethiopia stretch to Lake Tana, the source of the Blue Nile. The majority of Ethiopia's people live in rural areas, either in villages or isolated homesteads, much as their ancestors did.

# LAND AND PEOPLE

Ethiopia is a land of rugged mountains with a high, fertile plateau that stretches across about two-thirds of the country. The plateau covers much of the western and central regions. The Ethiopian Plateau, which lies between 6,000 and 10,000 feet (1,800 and 3,000 meters) above sea level, is crossed by deep river gorges and mountain ranges that rise more than 14,000 feet (4,300 meters).

Most Ethiopians live on the plateau, which has the best farmland in the country and usually receives more than 40 inches (102 centimeters) of rain a year. However, severe droughts occur from time to time, sometimes causing famine.

The plateau is split by the Great Rift Valley, which runs north and south through eastern Africa. A series of lakes extends through the valley in southern Ethiopia, but the country's largest lake, Tana, lies in the northwest.

Religious leaders of the Ethiopian Orthodox Church wear magnificent robes to mark their rank. About 40 percent of Ethiopians are Christians. Most of the rest are Muslim.

The Blue Nile—called *Abay* in Ethiopia—flows out of Lake Tana for 20 miles (32 kilometers) before spilling over the spectacular Tisissat Falls. Eventually, the Blue Nile joins the White Nile in Sudan, west of Ethiopia, where the two branches form the Nile River.

The Ethiopian Plateau slopes downward in all directions toward lowland regions. The lowlands are thinly populated because of the hot, dry climate and because the soil is poor for farming.

Grasslands cover much of the plateau, as well as many lowland areas to the south and east. The grasslands are home to such wild animals as antelope, lions, elephants, and rhinoceroses. Forests cover part of the southwest.

## Ethnicity and language

The Ethiopian government classifies its people into groups based on the primary language they speak. There are about 80 languages spoken in Ethiopia. Major ethnic and language groups include the Oromo, Amhara, and Tigrayan peoples.

The Oromo, who inhabit nearly every part of the country, make up the largest ethnic and language group. They speak an African language related to Somali called Oromifa. The

An ancient church cut from solid rock is a reminder of Ethiopia's rich history. Much of the country's art—especially early paintings and writing—is related to the Ethiopian Orthodox Church, a Christian faith related to the Coptic, Greek, and Russian Orthodox churches.

## FAMINE

The devastating droughts that hit Ethiopia in the 1970's and 1980's led to widespread famine, and thousands of Ethiopians died of starvation, malnutrition, or disease. Many crowded into refugee camps where food from other countries was distributed. The United Nations estimated that in 1988 about 7 million Ethiopians were dependent on such foreign aid, but the country's military government placed more importance on fighting rebels in northeastern Ethiopia than in aiding famine victims. The government ordered international relief agencies to leave Eritrea and Tigray, even though those areas were hardest hit by famine. Other countries charged that the Ethiopian government was using starvation as a way to win the war against the rebels.

Red Cross camps were crowded with hungry refugees from the Ethiopian highlands during the mid-1980's, when famine gripped the country. Ethiopia continues to depend on emergency aid to feed its people.

Amhara live in the northern plateau and speak Amharic—the nation's official language. The Tigrayan peoples of the same region speak Tigrinya. Amharic and Tigrinya belong to the same language family as Arabic and Hebrew.

Ethiopia is also home to such ethnic groups as the Somali, who live in the southeast, and the Afar, who live in the east and northeast. In addition to their own language, many Ethiopians speak English or another Ethiopian or European language. Ge'ez, an ancient Ethiopian language, is still used in Ethiopian Orthodox religious ceremonies.

## Modern life

Nearly 85 percent of the Ethiopian people live in the countryside. Poverty is widespread in these rural areas, and each year, large numbers of Ethiopians move to urban areas hoping to find work. Schools, medical care, and such modern conveniences as electricity are also more available in the cities.

Ethiopia's cities have many modern buildings, and several skyscrapers rise above Addis Ababa, the capital. Many rural Ethiopians live in traditional round houses. These homes have walls made of wooden frames plastered with mud, and they have cone-shaped roofs of thatched straw or metal sheeting. In Tigray areas, many Ethiopians live in rectangular stone houses.

Rural Ethiopians often wear a white cotton cloth called a *shamma*. Men wear a shamma over a shirt, and women wear it over a dress. In the south, people traditionally wear clothing made of leather, or they may use a colorful cloth as a shawl and a waist wrap. Many people in the towns and cities wear clothing similar to that worn in Europe and North America.

An Afar woman prepares injera, a slightly sour pancake-shaped bread eaten by many Ethiopians. They often use injera to scoop up portions of wat, a spicy stew.

A dependency of the United Kingdom, the Falkland Islands lie about 320 miles (515 kilometers) east of the southern coast of Argentina. They form the southernmost part of the British Empire outside the British Antarctic Territory.

## History and government

English explorer John Davis sighted the Falklands in 1592. British captain John Strong landed on the islands in 1690 and named them for Viscount Falkland, the British treasurer of the navy.

Because they lie on the important sea route around the tip of South America, the Falklands became an important stop for vessels rounding Cape Horn. Sailing vessels could drop anchor in the superb natural harbor of Stanley, take on supplies, and repair damage caused by the southern gales. These advantages led France, Spain, and Argentina to also lay claim to the islands, but the British established control there in 1833.

The Falklands are now an important British base. Nevertheless, Argentina has continued to claim the islands, which it calls Islas Malvinas.

In April 1982, Argentine troops invaded and occupied the islands. The United Kingdom responded by sending troops, ships, and planes to the Falklands. Air, sea, and land battles then broke out between Argentina and the British. The Argentine forces surrendered in June 1982, the economy considerably damaged.

In 1995, Argentina and the United Kingdom signed an agreement to begin drilling operations in the waters around the Falkland Islands. Drilling began in 1998.

## Environment and people

A governor, aided by an executive and a legislative council, rules the British dependency. The government provides schools, which children must attend. Traveling teachers instruct children in isolated settlements.

The British dependency includes two large islands, East and West Falkland, and hundreds of smaller ones. East Falkland covers 2,550 square miles (6,605 square kilometers), and West Falkland covers 1,750 square miles (4,533 square kilometers). Together, the islands have a coastline of 800 miles (1,288 kilometers). A vast area of islands and ocean became dependencies of the Falkland Islands Colony in 1908. The principal islands included South Georgia, South Orkney, South Shetland, and South Sandwich. The South Orkney and South Shetland island groups became part of the British Antarctic Territory in 1962. The territory includes the area south of 60° south latitude, between 20° and 80° west longitude.

The Falkland Islands lie in the South Atlantic, off the southern tip of South America. Although Argentina claims them, the British established rule there in 1833, and most of the inhabitants are of British origin.

Stanley, the capital and main settlement of the Falklands, sits on a fine, natural harbor. Strong winds limit the growth of trees on the islands, which consist mainly of rolling pastures.

Small outlying communities dot the landscape of the Falklands. While sheep farming is still a major activity, the sale of fishing licenses to foreign fishing fleets provides the main source of income.

A bleached whale bone on Sea Lion Island is a remnant of the days when whaling ships sailed the South Atlantic. American whaling flourished on all the world's oceans throughout the first half of the 1800's. During this period, thousands of whales were killed each year.

New Island, off West Falkland, is home to large colonies of penguins, including rockhoppers. The wildlife of the islands includes seals and many varieties of sea birds. Small islands set aside as nature preserves help protect the animals.

The landscape of the Falkland Islands resembles that of Scotland or Ireland, with many small bays and inlets along the coast, and low, rolling, grass-covered hills in the interior. Small streams wind through the shallow valleys, where many wildflowers grow. The temperature seldom falls below 18° F (–8° C) in winter. In summer, temperatures rarely rise above 77° F (25° C). Annual rainfall averages about 25 inches (65 centimeters), and snow falls occasionally in winter. Strong winds limit the growth of trees on the islands.

The Falklands are a major breeding ground for dozens of species of birds, including many varieties of penguins. The coastline also supports large populations of marine mammals, including certain types of porpoises, dolphins, seals, and sea lions.

Most of the approximately 3,000 inhabitants are of British origin. About three-fourths of the people live in Stanley, the capital and chief town, which lies on East Falkland Island.

Raising sheep was once the main economic activity of the islanders. While many of the islanders still raise sheep and export wool, the Falkland Islands' main source of income comes from the sale of fishing licenses to foreign fishing fleets. The sale of postage stamps and coins, primarily to collectors, also contributes to the economy.

# THE FAROE ISLANDS

A group of islands and reefs known as the Faroe Islands occupies a remote area in the North Atlantic Ocean between Iceland, Norway, and Scotland. The Faroes consist of 18 inhabited islands and a number of smaller islets and reefs. They cover an area of 540 square miles (1,399 square kilometers). The major islands are Streymoy, Eysturoy, Vágar, Sudhuroy, and Sandoy.

## Environment

The high and rugged landscape of the Faroes, with their steep and deeply indented coastlines, reflects the islands' violent origins. Millions of years ago, molten rock flowed up out of cracks in Earth's crust and spread out over the seabed. The molten rock solidified in layers of *basalt* (black volcanic rock) and red ash.

Later, Ice Age glaciers shaped the rock by gouging out steep-sided valleys and smoothing the basalt. The Faroes took their present shape about 10,000 years ago, when the Ice Age ended, causing the glaciers to disappear and the sea to flood into the valleys. Today, the Faroes' angular landscape consists of steep cliffs (known locally as "hammers"), numerous tiny rivers and waterfalls, chains of valleys dotted with lakes, and broad, green meadows.

Strong, gusting winds, raging storms, and crashing breakers add to the Faroes' rugged, desolate environment, while treacherous currents along the shores of the islands make navigation difficult. But, despite the Faroes' position near the Arctic Circle, the flow of the warm Gulf Stream guarantees surprisingly mild weather, with frequent fog and rain and relatively ice-free harbors. During the brief but beautiful summer, the midnight sun barely sinks below the horizon, creating glistening points of light on the deep waters of the fiords.

Faroese farm workers gather hay on a slope above a fishing village. The hay is used for animal feed during the winter. Although the islanders raise sheep for their wool and meat, livestock production no longer has the economic importance it once did.

## History

The inhabitants of the Faroes are hardy people of Norse origin. They fish and raise sheep for a living, and they also sell the eggs and feathers of the many sea birds that nest on the cliffs. Descended from the Vikings, the present-day islanders are peace-loving people who are proud of their Norse heritage.

Among the first settlers to arrive in the Faroes were Irish monks who settled in this remote corner of the world about 700. About 100 years later, Norse Vikings colonized the islands. Many of the settlers were refugees escaping the hard rule of Harold I (Fairhair), the brutal leader who unified Norway around 900. Others were drawn to the islands by a sense of adventure.

Norway ruled the Faroes from the 800's until 1380, when the islands came under the control of Denmark. Danish then replaced the Faroese tongue of the islanders as the official language, but the people held on to their national pride and cultural identity during the years of Danish rule. Faroese—which is related to Icelandic and Norwegian, and preserves many features of the Old Norse tongue—was restored as the territory's official language in 1948.

In 1948, the Faroes became a self-governing part of the kingdom of Denmark, with their own *Lagting* (parliament), flag, and currency. The Lagting sends representatives to the Danish parliament in Copenhagen.

## Way of life

Today, about 49,000 Faroese live in the islands—most of them in Tórshavn, which is the capital of the Faroes and also the economic and cultural center. Fishing and fish processing are the most important economic activities, with fish and fish products making up the majority of the islands' exports. Through their Lagting, the Faroese have established their own policy on fishing, choosing not to join the European Union (EU), which sets fish quotas.

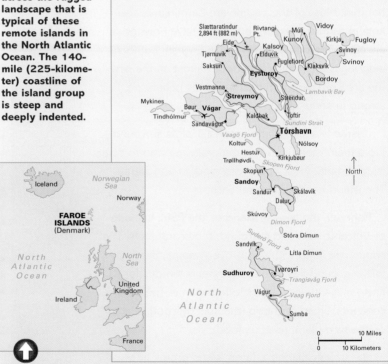

A Faroese village spreads out across the rugged landscape that is typical of these remote islands in the North Atlantic Ocean. The 140-mile (225-kilometer) coastline of the island group is steep and deeply indented.

The Faroe Islands are situated in the North Atlantic Ocean between Iceland, the United Kingdom, and Norway. The island group is a self-governing territory of Denmark. The national symbol of the Faroes is a ram, a reminder that raising sheep was once the major economic activity of the islands.

# FEDERATED STATES OF MICRONESIA

The Federated States of Micronesia (FSM), an independent country since 1986, consists of 607 islands just north of the equator. The FSM and Palau make up the Caroline Islands, a group of more than 930 islands scattered over a broad expanse of the Pacific Ocean.

## Land and people

The FSM lies between the Marshall Islands and the Philippines. The islands are scattered across about 1 million square miles (2.5 million square kilometers) of ocean, but the combined land area of the islands is only 271 square miles (702 square kilometers). The capital city is Palikir, located on Pohnpei.

About 100 of the 607 islands in the FSM are inhabited. Most of the people are Carolinians, also known as Micronesians, but some have Polynesian ancestry. The official language is English, but local languages are commonly used. Almost all the people are Christians, but local religious beliefs are strong in some areas.

Kosrae Island is made up of 15 islets surrounding a main island, but it is generally referred to as a single island. Most Kosraeans work in agriculture. The island has four natural harbors.

Pohnpei, the largest island of the FSM, is made of volcanic rock and surrounded by coral reefs. The shores are lined with mangrove swamps, but firm, fertile ground lies inland. Pohnpei is famous for its fine yams. Other crops include bananas, breadfruit, coconuts, limes, and *taro* (the starchy, rootlike stem of the taro plant, used as food).

The Chuuk Islands, a large island group, lie west of Pohnpei. About 48 of these islands are surrounded by a barrier coral reef that forms a lagoon 40 miles (64 kilometers) wide. Weno, the largest city in the FSM, lies on one of the Chuuk Islands.

## FACTS

| | |
|---|---|
| Official name: | Federated States of Micronesia |
| Capital: | Palikir |
| Terrain: | Islands vary geologically from high mountainous islands to low, coral atolls; volcanic outcroppings on Pohnpei, Kosrae, and Chuuk |
| Area: | 271 mi² (702 km²) |
| Climate: | Tropical; heavy year-round rainfall, especially in the eastern islands; located on southern edge of the typhoon belt with occasionally severe damage |
| Main rivers: | N/A |
| Highest elevation: | Totolom, 2,595 ft (791 m) |
| Lowest elevation: | Pacific Ocean, sea level |
| Form of government: | Republic |
| Head of state: | President |
| Head of government: | President |
| Administrative areas: | 4 states |
| Legislature: | Congress with 14 members |
| Court system: | Supreme Court |
| Armed forces: | U.S. is responsible for the Federated States of Micronesia's defense |
| National holiday: | Constitution Day - May 10 (1979) |
| Estimated 2010 population: | 113,000 |
| Population density: | 417 persons per mi² (161 per km²) |
| Population distribution: | 78% rural, 22% urban |
| Life expectancy in years: | Male, 68; female, 70 |
| Doctors per 1,000 people: | 0.6 |
| Birth rate per 1,000: | 24 |
| Death rate per 1,000: | 5 |
| Infant mortality: | 33 deaths per 1,000 live births |
| Age structure: | 0-14: 36%; 15-64: 60%; 65 and over: 4% |
| Internet users per 100 people: | 14 |
| Internet code: | .fm |
| Languages spoken: | English (official), Chuukese, Kosrean, Pohnpeian, Yapese, Ulithian, Woleaian, Nukuoro, Kapinga-marangi |
| Religions: | Roman Catholic 50%, Protestant 47%, other 3% |
| Currency: | United States dollar |
| Gross domestic product (GDP) in 2008: | $243 million U.S. |
| Real annual growth rate (2008): | N/A |
| GDP per capita (2008): | $2,165 U.S. |
| Goods exported: | Fish products, betel nuts, clothing, kava |
| Goods imported: | Electronics, food, machinery, motor vehicles |
| Trading partners: | Australia, Guam, Japan, United States |

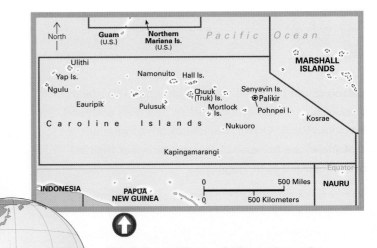

Young Pohnpei islanders participate in a formal ceremonial dance. Pohnpei is the largest island of the FSM.

The Federated States of Micronesia consists of Kosrae Island, Pohnpei Island, the Chuuk Islands, and the Yap Islands. The islands are scattered across about 1 million square miles (2.5 million square kilometers) of ocean.

The Yap Islands, separated by long, narrow channels, lie at the western end of the FSM.

The Yap Islands, which lie west of the Chuuks, include 4 large islands and 10 smaller islands separated by narrow channels and ringed by coral reefs. They are composed of ancient crystalline rocks and have a rugged surface. Most of the Yap Island people make a living by farming. Fishing is also important.

## History

The Yap Islands were among the first island groups settled in Micronesia. Archaeologists believe that people from Asia moved there thousands of years ago. Settlers arrived on Kosrae, Pohnpei, and the Chuuk Islands later. The ruins of Nan Madol on Pohnpei are the remnants of an advanced civilization dating from some time after the original Micronesians arrived from Asia. Mysterious structures made of huge blocks are all that remain of palaces, administrative buildings, houses, and places of worship.

In the 1500's, Spanish explorers were the first Europeans to reach the islands. Spain formally claimed them in 1885 and sold them to Germany in 1899. In 1905, the Yap Islands became internationally important as a cable station between the United States, the Netherlands Indies (now Indonesia), and Japan.

Japan captured the islands during World War I (1914–1918). After the war, the League of Nations gave them to Japan as mandates. During World War II

(1939–1945), U.S. forces captured some of the islands. In 1947, the United Nations made the United States trustee of the islands as part of the Trust Territory of the Pacific Islands.

In 1980, Kosrae, Pohnpei, the Chuuks, and the Yaps formed the Federated States of Micronesia in agreement with the United States. In 1986, these islands became an independent country. The United States provides economic aid and defends the islands in emergencies.

# FIJI

The island country of Fiji is made up of about 330 islands and about 500 more tiny atolls, islets, and reefs scattered in the South Pacific Ocean. The islands are considered part of Melanesia because of their location, but their culture is more like that of Polynesia.

Fiji has been called the "crossroads of the South Pacific" because it lies on major shipping routes and has several excellent harbors. Also, its airport is a busy terminal for planes flying the Pacific.

Fiji has about 877,000 people. The population consists of two major groups with different origins, languages, cultures, and religions. The native Fijians are Melanesians whose ancestors arrived in Fiji thousands of years ago, probably from Indonesia. The other group is made up of descendants of laborers brought from India between 1879 and 1916 to work on sugar plantations. A small number of Fijians are of Chinese, European, Micronesian, or Polynesian ancestry.

English is the official language of Fiji and is used in the schools. But Fijian and Hindi are also widely spoken. More than 85 percent of those from 6 to 13 years old attend school, although it is not required by law. Most Fijian and Indian youngsters attend separate schools. The University of the South Pacific in Suva, the country's only university, serves students from many of the Pacific Islands.

## Economy

Fiji's economy is based primarily on agriculture. Most Fijians grow such crops as sugar cane and coconuts. The country also exports timber. Gold is Fiji's chief mineral resource. Sugar cane is Fiji's principal cash crop, but weather conditions, especially drought, and unstable prices in international markets have made the nation's income from sugar cane increasingly unpredictable.

Since Fiji became independent in 1970, the government has encouraged tourism, which employs many islanders, and the development of manufacturing and forestry. It has also promoted the production of new crops to reduce Fiji's dependence on sugar cane and coconuts. In 1978 an Australian aid program was launched to turn hilly, undeveloped Fijian land into grazing land. In addition, Australia and Fiji expanded a jointly owned

## FACTS

| | |
|---|---|
| Official name: | Republic of the Fiji Islands |
| Capital: | Suva |
| Terrain: | Mostly mountains of volcanic origin |
| Area: | 7,078 mi² (18,333 km²) |
| Climate: | Tropical marine; only slight seasonal temperature variation |
| Main river: | Rewa |
| Highest elevation: | Mount Tomanivi 4,341 ft (1,323 m) |
| Lowest elevation: | Pacific Ocean, sea level |
| Form of government: | Republic |
| Head of state: | President |
| Head of government: | Prime minister |
| Administrative areas: | 4 divisions, 1 dependency |
| Legislature: | Parliament consisting of the Senate with 32 members and the House of Representatives with 71 members serving five-year terms |
| Court system: | Supreme Court |
| Armed forces: | 3,500 troops |
| National holiday: | Independence Day - Second Monday of October (1970) |
| Estimated 2010 population: | 877,000 |
| Population density: | 124 persons per mi² (48 per km²) |
| Population distribution: | 52% urban, 48% rural |
| Life expectancy in years: | Male, 67; female, 72 |
| Doctors per 1,000 people: | 0.5 |
| Birth rate per 1,000: | 22 |
| Death rate per 1,000: | 6 |
| Infant mortality: | 14 deaths per 1,000 live births |
| Age structure: | 0-14: 31%; 15-64: 64%; 65 and over: 5% |
| Internet users per 100 people: | 11 |
| Internet code: | .fj |
| Languages spoken: | English, Fijian, Hindustani (all official) |
| Religions: | Christian 64.5%, Hindu 27.9%, Muslim 6.3%, other 1.3% |
| Currency: | Fijian dollar |
| Gross domestic product (GDP) in 2008: | $3.63 billion U.S. |
| Real annual growth rate (2008): | 1.2% |
| GDP per capita (2008): | $4,148 U.S. |
| Goods exported: | Fish, gold, petroleum products, sugar, timber |
| Goods imported: | Food, machinery, motor vehicles, petroleum products |
| Trading partners: | Australia, New Zealand, Singapore, United States |

Fijian gold mine, thus increasing the nation's gold output. Over time, construction of new lumber mills expanded the timber industry. In the late 1900's, tourism overtook the sugar industry as the major source of foreign earnings.

## Government

Fiji was a British colony between 1874 and 1970. After the nation became independent, Fijians held more power in government than Indians. In 1987, a Fijian prime minister appointed a multiracial Cabinet. Military officers seized power in protest. Later in 1987, Fiji was returned to civilian rule.

In 1990, Fiji approved a new constitution, which established a two-house legislature. Elections for legislators took place in 1992. In 1997, the Constitution was amended to grant political power to all races. In 1999, Fiji elected its first Indian prime minister.

In May 2000, a group of rebels, claiming to represent the native Fijians, stormed Parliament and held the prime minister and most members of the Cabinet hostage. Military leaders took control of the country, setting up an interim government and revoking the 1997 Constitution. In November 2000, Fiji's high court ruled that the 1997 Constitution was still in effect. A new interim government was established in March 2001 with an ethnic Fijian as prime minister. The prime minister appointed a cabinet

A group of Fijian children reflects the multiracial nature of the islands.

that excluded representatives of parties dominated by ethnic Indians. In 2002, Fijian courts ruled that such parties must be included.

In 2006, Frank Bainimarama, Fiji's military chief, seized control of the government. The next year, he was sworn in as interim prime minister, and in 2009, he imposed martial law. Bainimarama lifted the state of martial law in January 2011 but imposed new emergency laws.

Workers harvest sugar cane by hand under the bright Fijian sun. Sugar cane is one of Fiji's chief agricultural products.

Fiji is often called the "crossroads of the South Pacific" due to its excellent location, harbors, and airport.

# ENVIRONMENT

Fiji has a total land area of 7,078 square miles (18,333 square kilometers). The Fijian island of Viti Levu (Big Fiji) accounts for about half the land area of the country. Suva, Fiji's capital and largest city, lies on Viti Levu's southern coast. A second island, Vanua Levu (Big Land) occupies about a third of Fiji's land area. Only about a hundred of the more than 800 islands of Fiji are suitable for human habitation. Many of the other islands are little more than piles of sand on coral reefs.

Cool winds make Fiji's tropical climate relatively comfortable. Temperatures range from about 60° to 90° F (16° to 32° C). Heavy rains and tropical storms occur frequently between November and April.

## Volcanic islands

Most of the islands of Fiji consist of lava built up from the ocean floor by eruptions of oceanic volcanoes. The volcanoes are no longer active, but the heat trapped under the islands now causes hot springs to flow, for example, on Vanua Levu.

Building a house in Fiji is a traditional skill handed down through generations. The people use local materials such as bamboo and reeds for their homes and mangrove wood for their cooking fires.

An industrial area near Fiji's capital of Suva produces cement, beer, and other goods. Suva is Fiji's chief port and commercial center.

The larger volcanic islands have fertile soil that has developed from the weathered volcanic rock. The larger islands also have high volcanic peaks, rolling hills, rivers, and grasslands. The mountainous interiors of these islands form a barrier that keeps most moisture of the southeasterly trade winds on the eastern side, leaving the west drier and typically covered by grasslands. On the southeastern slopes of the islands, where rainfall is heaviest, dense rain forests thrive. These tropical rain forests cover more than half the total area of Fiji.

The view from Mana Island shows the country's dense vegetation. Tropical rain forests cover more than half of Fiji. The larger islands also have high volcanic peaks.

## Wildlife

Mangrove swamps cover the coastal flats and help to build up the Fiji islands. Mangrove trees usually grow in places by quiet ocean water. Their thousands of stiltlike roots catch silt, which piles up to form new land. The roots also form a breeding place for many fish and other creatures. The clear seas around the coral reefs support a wide variety of marine life.

Over the generations, Fijians in certain areas have developed the custom of calling up sea creatures from the deep. For example, on Kadavu and Koro, people chant a traditional song to lure turtles into shallow water. On Vatulele Island, people summon red prawns with a certain call. And in Lakeba, an island in the Lau group, people still practice the art of calling sharks.

In contrast to the richness of marine life, few land animals inhabit the islands. Originally, volcanic and coral islands have no land animals or plants, but these islands become inhabited by birds that fly across the sea and by other animals that swim to the islands. Some animals and insects are carried to the islands on logs or other debris.

In Fiji, the only native mammals are a few species of bats and a small Polynesian rat. Pigs and dogs came to the islands with the first settlers, but all other mammals have arrived since the Europeans began to settle there in the 1800's.

Cleared for use as farmland, many inland parts of the Fiji Islands now produce sugar cane, rice, and other crops. Coconuts are an important cash crop, and pine forests have also been planted to provide valuable timber for export.

Coral reefs—limestone formations composed of tiny sea animals and plants and their remains—surround nearly all the islands. Fiji also has coral islands, which consist chiefly of coral reef material piled up by ocean waves and winds. Coral islands tend to be flat with sandy beaches and lie only a few feet above sea level. The larger coral islands often have rich vegetation, including vines, grasses, broadleafed trees, and coconut palms.

Fiji's fertile soils support the growth of sugar, coconuts, and bananas. Fiji also has small deposits of gold and manganese.

# HISTORY AND PEOPLE

In 1643, Abel Tasman, a Dutch navigator, became the first European to see Fiji. Captain James Cook, a British explorer, visited Vatoa, one of the southern islands, in 1774. During the 1800's, traders, Methodist missionaries, and escaped Australian convicts came to visit or settle there.

Various tribes fought one another, and Fijians had a reputation as the most savage and warlike of all the Melanesian peoples. In 1871, however, a chief named Cakobau gained power and brought peace to Fiji. To protect his country from outside interference, Cakobau asked the United Kingdom to make Fiji a crown colony. The United Kingdom did so in 1874.

## Indians and Fijians

Soon after Fiji became a crown colony, the colonial government decided to import about 60,000 laborers from India to work on Fiji's plantations. The traditional religious and caste systems of the Indians were ignored, and they often had to work under very harsh conditions. They were looked down upon by both the Fijians and the Europeans, and they struggled for political and social recognition.

Fiji became an independent nation at its own request in 1970. Throughout the years, the differences between the native Fijians and the Indians have led to racial violence and political upheaval. Although Indians control most of Fiji's economy, native Fijians traditionally have held more power in the government. In April 1987, however, an Indian-backed coalition led by Timoci Bavadra won a majority in Parliament. Bavadra became prime minister and appointed Indians to a majority of the Cabinet posts.

Many Fijians resented this increase of Indian political power, and military officers overthrew the government. The new military leader, Colonel Sitiveni Rabuka, abolished the Constitution, named himself head of state and government, and declared the right of Fijians to govern the nation. In December 1987, Rabuka appointed a president and returned Fiji to civilian rule.

A spectacular tropical sunset magnifies Fiji's beauty. Fiji lies south of the equator. Although it has a hot, tropical climate, cool winds make it relatively comfortable.

Fire walking is a tradition among men of the Sawau tribe from the island of Beqa. By tradition, the men must avoid women and coconuts for two weeks before the ceremony. They believe that if they do not, they may burn their feet.

In July 1990, Fiji approved a new constitution, establishing a two-house legislature. In 1997, the Constitution was amended to grant political power to all races. And in 1999, Mahendra Chaudhry became Fiji's first prime minister of Indian descent.

In 2000, a group of rebels stormed Parliament, seized Chaudhry and most members of the Cabinet, and held them hostage. The rebels claimed to represent the interests of native Fijians. Chaudhry was eventually removed from office, and military leaders took control of the country. The military rulers set up an interim government and revoked the 1997 Constitution, though Fiji's high court later ruled that the 1997 Constitution was still in

force. Laisenia Qarase, an ethnic Fijian, was named prime minister in 2000.

In 2006, Frank Bainimarama, Fiji's military chief, seized control of the government. In 2007, he restored power to the president, who swore in Bainimarama as interim prime minister. Bainimarama repeatedly refused international pressure to set a date for new elections. In 2009, the Pacific Islands Forum and the Commonwealth of Nations both suspended Fiji's membership.

## Modern life

Many Fijians of Indian descent still work in the cane fields, but others have become shopkeepers or business people. The Indian women wear *saris,* the traditional dress of Indian women. Most Indians are Hindus or Muslims.

About half of the native Fijians live in rural areas. Chiefs still play an important part in the affairs of many villages.

On most islands, the villagers celebrate such occasions as births and marriages by traditional feasting, dancing, and singing. Important festivals include the ceremonial drinking of *kava,* a drink made from pepper plants. The men often wear a cloth skirt called a *lava-lava* or *sulu,* and some of the women wear long, loose-fitting cotton dresses called *muumuus.*

Toberua Island, off the eastern coast of Viti Levu, has a popular tourist resort and the calm, peaceful beauty most people associate with a tropical paradise.

Catamarans, raftlike boats with two hulls, allow visitors to enjoy Fiji's beautiful coastal waters. The modern version has been adapted from a traditional Polynesian outrigger design developed thousands of years ago.

Finland, a country famous for its scenic beauty, lies in northern Europe north of the Baltic Sea and east of Sweden. About one-third of the country lies within the Arctic Circle in an area known as the Land of the Midnight Sun. In this region, the sun shines 24 hours a day for long periods each summer. During the endless winter nights, the *aurora borealis* (northern lights) fill the sky with colorful curtains of light.

Finland is a prosperous nation, largely because of the huge forests that cover most of the land. As a result of the nation's thriving economy, the Finnish people enjoy a high standard of living.

The Finns differ from the Scandinavians and the Slavs—the other peoples of northern and eastern Europe—in language and culture. Although Finland was dominated for centuries by its powerful neighbors, Sweden to the west and Russia to the east, the Finns never lost their national identity. Their rich folk culture is reflected today in the country's crafts, literature, music, and painting.

The Finns' contribution to world culture includes the *Kalevala,* a huge collection of song-poems and chants of Finnish peasants. The *Kalevala,* which became the Finnish national epic, inspired the works of other Finnish masters, such as the symphonic poems of Jean Sibelius and the paintings of artist Akseli Gallen-Kallela.

## The earliest Finns

The Sami, who now make up a small minority of the Finnish population, were Finland's earliest known inhabitants. The ancestors of the present-day Finns began to move into the country from the south shores of the Gulf of Finland thousands of years ago.

In the A.D. 1000's, both Sweden and Russia began a struggle for control of Finland. By the 1200's, Sweden was the dominant power, and Swedish was the nation's official language. Lutheranism became the official religion about 1540.

Russia came to fear Swedish expansion and sought to protect its western border by regaining Finland. From the 1500's to the 1700's, Russia fought a series of wars with Sweden. In 1808, Russia invaded Finland, and it finally conquered the country in 1809. Finland was made an independent grand duchy, with Russia's czar as grand duke. The duchy had local self-rule based on government systems developed while Finland was under Swedish control.

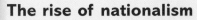

# FINLAND

## The rise of nationalism

In 1899, Czar Nicholas II took away most of Finland's powers of self-rule and began an attempt to force the Finns to accept Russian government and culture. But the Finns opposed the czar's actions, and resistance to Russian control reached a peak in 1905 with a six-day nationwide strike. The strike forced the czar to restore much of Finland's self-government, but he continued his efforts to "Russianize" the people.

In 1917, when a revolution in Russia overthrew the czar, Finland declared its independence. However, a civil war broke out in Finland between the socialist Red Guard, aided by the Russian Communists, and the antisocialist White Guard, assisted by German troops. The war ended in a White victory in May 1918. In 1919, Finland adopted a republican constitution.

The new country's relations with Sweden and Russia remained unsettled throughout the early 1900's. Finland and Sweden quarreled over possession of the Aland Islands until 1921, when the League of Nations awarded the islands to Finland. Disputes with Russia centered on the eastern Finnish region of Karelia. In November 1939, Soviet troops invaded Finland, and the Finns were forced to admit defeat after the 15-week "Winter War." A peace treaty signed in March 1940 gave southern Karelia—about 10 percent of Finland's total land area—to the Soviet Union.

During World War II (1939-1945), Finland allowed Nazi Germany to move troops through its northern regions to attack the Soviet Union, and the Soviets then bombed Finland. As the Germans retreated, they burned towns, villages, and forests behind them.

After the mid-1900's, the Finnish government kept a policy of neutrality in international politics. Finland eventually developed close economic and cultural ties with Denmark, Norway, Sweden, and Russia.

223

# FINLAND TODAY

After World War II, which caused a devastating loss of life and property in Finland, the country's leaders adopted a policy of neutrality in international politics. Finland also developed close economic and cultural ties with Russia and the countries of Scandinavia. On Jan. 1, 1995, Finland became a member of the European Union (EU). In 1999, Finland joined the eurozone, the group of EU nations that use the euro as a common currency.

Finland is a democratic republic, with a parliament called the *Eduskunta*. Parliamentary elections are based on a system of *proportional representation,* which means that the number of seats gained by a political party depends upon a share of the total number of votes cast. Under this system, it is often difficult for any single political party to establish an overall majority. Thus, Finland is generally governed by a *coalition,* in which several parties join together to form a government.

Finland's president is the country's head of state and chief executive. The president is elected to a six-year term by the people. The president appoints the prime minister, who is head of government. The prime minister forms a Cabinet made up of members of various political parties. Leading political parties of Finland include the Center Party, the Green Union, the National Coalition Party, and the Swedish People's Party. In 2000, Finland elected its first woman president, Tarja Halonen.

Like Sweden and Norway, Finland has an extensive welfare system, which provides the people with many services. Since the 1920's, maternity and child welfare centers have given free health care to pregnant women, mothers, and children. In 1939, Finland established an old-age and disability insurance program. Since 1948, families have received an allowance when they have a new baby, as well as a yearly allowance for each child under 16.

Finnish children are required to attend *basic school* for nine years, beginning at the age of 7. Almost all

## FACTS

| | |
|---|---|
| Official name: | Suomen Tasavalta (Republic of Finland) |
| Capital: | Helsinki |
| Terrain: | Mostly low, flat to rolling plains interspersed with lakes and low hills |
| Area: | 130,559 mi² (338,415 km²) |
| Climate: | Cold temperate; potentially subarctic, but comparatively mild because of moderating influence of the North Atlantic Current, Baltic Sea, and more than 60,000 lakes |
| Main rivers: | Kemijoki, Ounasjoki, Muonio |
| Highest elevation: | Mount Haltia, 4,344 ft (1,324 m) |
| Lowest elevation: | Baltic Sea, sea level |
| Form of government: | Democratic republic |
| Head of state: | President |
| Head of government: | Prime minister |
| Administrative areas: | 6 laanit (provinces) |
| Legislature: | Eduskunta (Parliament) with 200 members serving four-year terms |
| Court system: | Korkein Oikeus (Supreme Court) |
| Armed forces: | 29,300 troops |
| National holiday: | Independence Day - December 6 (1917) |
| Estimated 2010 population: | 5,323,000 |
| Population density: | 41 persons per mi² (16 per km²) |
| Population distribution: | 63% urban, 37% rural |
| Life expectancy in years: | Male, 76; female, 83 |
| Doctors per 1,000 people: | 3.3 |
| Birth rate per 1,000: | 11 |
| Death rate per 1,000: | 9 |
| Infant mortality: | 3 deaths per 1,000 live births |
| Age structure: | 0-14: 17%; 15-64: 66%; 65 and over: 17% |
| Internet users per 100 people: | 79 |
| Internet code: | .fi |
| Languages spoken: | Finnish, Swedish, Sami, Russian |
| Religions: | Lutheran Church of Finland 82.5%, Orthodox Church 1.1%, other 16.4% |
| Currency: | Euro |
| Gross domestic product (GDP) in 2008: | $273.98 billion U.S. |
| Real annual growth rate (2008): | 1.5% |
| GDP per capita (2008): | $51,841 U.S. |
| Goods exported: | Chemicals, electronics, machinery, paper and pulp, transportation equipment, wood and wood products |
| Goods imported: | Food, iron and steel, machinery, petroleum and petroleum products, transportation equipment |
| Trading partners: | China, Germany, Russia, Sweden, United Kingdom, United States |

Finland is noted for the scenic beauty of its sparkling waters, its forests of pine, spruce, and birch, and its many islands. Finland is often called the land of a thousand lakes.

A statue of Finnish soldier and statesman Carl Gustaf Mannerheim stands in front of Helsinki's parliament building. Mannerheim served as Finland's president between 1944 and 1946.

students attend public school, where they receive one free meal a day, as well as free books and medical and dental care. After basic school, students may choose an *upper secondary school,* which emphasizes academic subjects, or a *vocational school,* which offers training in skilled manual work. Vocational institutes and universities offer higher-education programs.

More than 90 percent of Finland's people are Finnish by descent, and most of the rest are Swedish. The majority of the people are tall, with fair skin, blue or gray eyes, and blond or light brown hair. Finnish and Swedish are both official languages. Most of the people speak Finnish, and many speak Swedish. Most of the Swedish-speaking people live on the south and west coasts and on the offshore Aland Islands.

The majority of Finland's people live in the south, and more than 60 percent live in cities and towns. In urban areas, most Finns live in apartments, while people in rural areas live in single-family homes on farms or in villages.

# ENVIRONMENT

Finland stretches about 640 miles (1,030 kilometers) from the Arctic Circle in the north to the Baltic Sea in the south. Much of the country consists of a low-lying plateau broken by small hills and valleys and low ridges and hollows.

The land rises gradually from south-southwest to north-northeast, averaging only about 400 to 600 feet (120 to 180 meters) above sea level. It is covered mainly by dense forests of pine, spruce, and birch trees, and about 60,000 lakes scattered throughout the countryside.

Finland's present landscape was formed more than a million years ago, when continental glaciers covered the land, gouging the surface as they advanced. When the glaciers receded about 10,000 years ago, they left deposits of rocks that formed *end moraines* or *terminal moraines*—irregular ridges with hills and hollows.

Finland enjoys a surprisingly mild climate for its northern latitude, largely because of the warm ocean current known as the Gulf Stream. The country's many lakes, as well as the gulfs of Bothnia and Finland, also help give the country a warmer climate. However, winters are long and often harsh. Snow covers southern Finland from November to April, while northern Finland is snowbound from October to April.

A lake near Kuopio, in the heart of Finland's Lake District, reflects the natural beauty of central Finland's landscape, with its interplay of lakes, islands, inlets, and forestland.

## Southern land regions

Finland's four main land regions are the Coastal Lowlands, the Coastal Islands, the Lake District, and the Upland District. The Coastal Lowlands, a region of many small lakes, lie along the Gulf of Bothnia and the Gulf of Finland. The lowlands have less forested land and a milder climate than the Lake and Upland districts, and they rank as Finland's most densely populated area.

The Coastal Islands consist of thousands of islands in the Gulf of Bothnia and the Gulf of Finland. The majority are small, uninhabited islands with thin, stony soil that cannot support much plant life. However, many kinds of plants grow on a few of the larger islands. Some islands are inhabited by people who fish for a living, but most are used as summer recreation areas.

The Lake District extends across central Finland north and east of the Coastal Lowlands. Island-dotted lakes, connected by narrow channels or short rivers, cover about half the total area of the district. Finland's largest lake, Saimaa, covers about 680 square miles (1,760 square kilometers) in the southeastern part of the region.

A trapper's hut, with a motorized sled parked outside, stands alone in the wilderness of the Upland District. Few people live in this region, where winter temperatures sometimes drop as low as −22° F (−30° C).

## The Upland District

The Upland District, Finland's northernmost region, covers about 40 percent of the country and ranks as the nation's most thinly populated region. The district has a harsher climate and less fertile soil than other regions. Hilly areas are separated by swamps and marshlands, and in the northernmost parts, the land is covered by frozen, treeless tundra.

The district's marshlands are the habitat of Finland's largest animal—the moose. This magnificent creature, a member of the deer family, has been reduced in numbers by hunting and deforestation. However, today moose are carefully protected, and their numbers are increasing.

Finland's only mountains are located in the extreme northwest of the Upland District, where the country's tallest peak, Mount Haltia, rises 4,344 feet (1,324 meters) along the Norwegian border. The Kemijoki, the longest river in Finland, begins in the Upland District near the Russian border and winds 340 miles (547 kilometers) southwestward to the Gulf of Bothnia.

## THE FINNISH SAUNA

A tradition in Finland for more than 1,000 years, the dry-heat bath known as the sauna is used for cleansing and relaxation. In a sauna, bathers sit or lie on wooden benches, perspiring freely from the warmth created by stones heated on top of a furnace. The room is made even hotter when water is thrown on the stones, causing steam (löly) to rise. Bathers also beat their bodies gently with birch twigs (vihta) to stimulate circulation. Some people end a sauna bath with a roll in the snow, but most settle for a cold shower. Then they lie down to rest until their body returns to normal temperature.

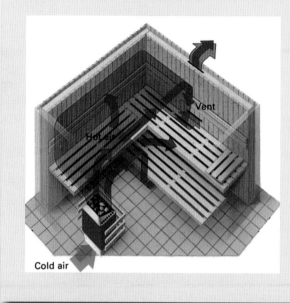

The Aland Islands are the most important of Finland's Coastal Islands. Most of the approximately 6,500 Aland Islands are inhabited, mostly by Swedish-speaking people. The main island in this group, also called Aland, is Finland's largest island.

# HELSINKI

Helsinki, Finland's capital and largest city, is a thriving seaport on the Gulf of Finland, on the country's southern coast. Helsinki covers a peninsula and several islands. Founded in 1550 by King Gustavus I of Sweden, the city is one of Finland's chief ports as well as the nation's commercial and cultural center. About one-fifth of the Finnish people live in Helsinki and its suburbs. The city is home to Finland's main university as well as to theaters, art galleries, and museums.

During its early days, Helsinki suffered a series of disasters, including fires, plagues, war, and famine. Between the 1500's and the 1700's, Helsinki was almost destroyed twice in the fighting between Russia and Sweden over control of Finland. Later, the Great Fire of 1808 burned more than two-thirds of the city to the ground.

In 1809, the city passed from Swedish to Russian control, and in 1812, Czar Nicholas II made Helsinki the capital of Finland. The royal order also included a large-scale program of city planning and construction, designed to reflect the city's new political importance.

The German-born architect Carl Ludwig Engel was directed to create a city that combined innovative architecture with a spacious layout of broad avenues and numerous parks. Present-day Helsinki's reputation as a showpiece of modern architecture is largely due to the efforts begun by Engel and his partner, Johan Albrekt Ehrenström.

## Architectural wonders

Engel's inspiration was the neoclassical architecture of St. Petersburg in Russia. His impressive works can be seen in Helsinki's imposing Senate Square, surrounded by the Government Palace, the University of Helsinki, and the Lutheran Cathedral with its gleaming white columns and shining domes.

Senate Square lies in the heart of Helsinki's old section. The square is dominated by Lutheran Cathedral, designed in the Neoclassical style. A statue of the Russian Czar Alexander II stands in the square. Finland was under Russian control for the most of the 1880's and early 1900's.

Helsinki's main street typifies the city's spacious atmosphere, with the sea never far away. Helsinki is a city of scenic bays and broad, treelined streets. The contrast of old and new architecture adds to the city's unique charm.

The tradition of dazzling architectural design begun by Engel continued in Aleksander Gornostayev's Uspensky Cathedral, completed in 1868. This cathedral is a vast Byzantine structure of red brick and golden cupolas, with a lavishly decorated interior. Standing in dramatic contrast to the work of Engel and Gornostayev is the pink granite and art deco styling of Eliel Saarinen's railway station, which in 1914 seemed shockingly innovative.

Almost every street in Helsinki features works by distinguished Finnish architects, including Alvar Aalto, Herman Gesellius, and Saarinen. Notable among the most recent examples of Finnish architecture is Aalto's white marble and limestone Finlandia Hall, which opened in 1971.

## A city of the sea

Helsinki is surrounded by the sea on three sides, and fishing boats set sail from its harbors every morning. Their catch of Baltic herrings (silakka) is sold directly from the boats at South Harbor quay.

Near the fish stalls stands the bronze statue of the sea maiden Havis Amanda. The statue is the center of traditional May Day Eve festivities. On this night, students scramble into the protective moat surrounding the statue and crown Havis Amanda with their white caps.

Across from the statue are the flower, fruit, and vegetable stalls of Market Square, Helsinki's morning market. In the summer months, the market reopens in the late afternoon, and street vendors sell local handicrafts to tourists. Also in the summer months, the fruit stalls are laden with wild berries ripened by the midnight sun—cranberries, whortleberries, rowanberries, and the delicately flavored Arctic cloudberry known as *suomuurain*.

Finlandia Hall—a marble concert hall on Helsinki's Töölönlahti Bay—was designed by Finnish architect Alvar Aalto and completed in 1971. The tower of the National Museum rise in the background. The museum was completed in 1904 by Finland's other famous architect, Eliel Saarinen

# ECONOMY

Before the outbreak of World War II (1939–1945), the economy of Finland depended largely on agriculture, forestry, and fishing. The war, however, left the nation's economy in a desperate condition. About 10 percent of its productive capacity was lost to the Soviet Union, and the country was forced to take in about 400,000 war refugees.

From 1944 to 1952, Finland also had to pay the Soviet Union $300 million in war damages. The cost of these reparations, in addition to rising inflation and an increase in the Finnish population, severely burdened the economy. However, the demands of reparation payments also helped fuel industrial growth, and the country's metal-working industry developed rapidly.

By 1947, Finland's gross national product (GNP) had reached its prewar level, and since then, the economy has enjoyed consistent growth. The Finnish people have built a productive and diversified economy despite the country's relatively poor soil, harsh climate, and lack of mineral resources.

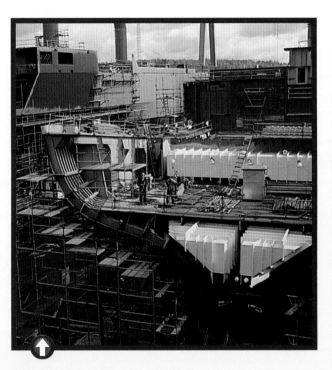

Turku, on the southwest coast, is known for its shipyards. The city is important to Finland's economy because its harbor is ice-free and more easily kept open in winter than the nation's other seaports.

Finland's prosperity depends heavily on international trade. The nation's most valuable exports are electronics, forestry, and metal-working products, particularly paper and paper products, ships and other transport equipment, and industrial machinery. Finland's largest company, Nokia, is a major manufacturer of cell phones. Finland belongs to the European Union (EU), an economic and political partnership of European countries. In 1999, Finland adopted the EU's common currency, the euro.

## Forestry and agriculture

Finland is the most heavily forested of all European countries. Its forests cover more than 70 percent of the land and remain its greatest natural resource. As a result, forestry plays a key role in Finland's economy. Pine, birch, and spruce trees make up most of the forests.

The Finnish government owns about a third of the nation's forests, chiefly in the north. The southern forests, owned mostly by private farmers, are the most productive because of the longer growing season and the cheap, efficient transport provided by the nation's network of lakes and waterways. Careful conservation and replanting have largely made up for the loss of southeastern Finland's Karelian forests to the Soviet Union after World War II.

Today, Finland is the world's leading producer of plywood, and the manufacture of paper, paperboard, and wood paneling are major industries. Prefabricated houses, which are erected in factory-made sections, are another major forest product.

Agriculture in Finland is limited by the relatively small proportion of suitable farmland. Most Finnish

Huge rafts of timber are floated down rivers and assembled on Finland's Lake Saimaa, the largest of Finland's extensive lake systems. Steamships travel through the lakes, stopping at towns along the way.

Papermaking ranks as one of Finland's most important economic activities. Paper production and other forest-based industries traditionally have been Finland's chief manufacturing industries.

farms lie in the south and west, and their average size is 77 acres (31 hectares). Nevertheless, Finland's farmers produce all the milk, eggs, meat, cereals, and bread grains needed by the people. Barley and oats are the main grain crops. Other crops include potatoes, sugar beets, and wheat.

## Manufacturing and energy

The metal-working industry has become as important to Finland's economy as its forest-based industries. The nation's metal-working industry produces electric motors and generators, farm equipment and machinery, and machinery used in the paper and lumber industries. Other manufactured products include chemicals, printed items, processed foods, and transportation equipment. Finnish shipyards, which build vessels suitable for cold northern waters, are noted for their powerful ice-breakers. The Finns also produce high-quality recreational yachts that are sold throughout the world.

Hydroelectric plants on the major rivers provide power for Finland's industries. During the 1970's and 1980's, Finland completed the construction of four nuclear power plants that now provide for much of the nation's energy needs.

A woman talks on her mobile phone in downtown Helsinki. Nokia, one of the world's largest mobile telephone producers, has its headquarters in Finland.

# FRANCE

High atop Montmartre, the tallest hill in Paris, stands the Basilique du Sacré Coeur (Basilica of the Sacred Heart). With its huge bell tower and onion-shaped dome, this gleaming white church is one of the city's most famous landmarks.

As an international capital of art, fashion, and learning, Paris ranks among the world's great cities. But Paris, the beautiful City of Light, as it is known the world over, is only one of France's treasures.

From the rocky, mist-covered cliffs of Normandy to the sunny beaches of Nice, France is a country of great diversity and charm. With its colorful apple orchards, dairy farms, and vineyards, the French countryside has a charm all its own. Historic *châteaux* (castles) still dominate the Loire Valley, and picturesque fishing villages line the Atlantic coast in the northwest.

With its fertile soil and mild climate, France is Europe's largest agricultural producer and a leading exporter of farm products. The country ranks as one of the world's leading manufacturing nations, with thriving automobile, chemical, and steel industries. In addition, France plays an important role in world politics and economic affairs.

France has a long and colorful history. Julius Caesar and his Roman soldiers conquered the region before the time of Christ. Then, after Rome fell, the Franks and other Germanic tribes invaded the region. France was named for the Franks. By the A.D. 800's, the mighty Charlemagne, king of the Franks, had built the area into a huge kingdom.

In 1792, during the French Revolution, France became one of the first nations to overthrow its king and set up a republic. A few years later, Napoleon Bonaparte seized power. He conquered much of Europe before he finally was defeated. During World Wars I and II, France was a bloody battleground for Allied armies and the invading German forces.

Throughout their history, the French people have been among the leaders of European culture. French styles in painting, music, drama, and other art forms have inspired countries throughout the Western world.

The French people are also noted for their *joie de vivre*—their celebrated enjoyment of life. They especially value good food and wine. Fine French cooking has been raised to an art form. Their relaxed way of life also includes the conversation and the company of friends at a sidewalk cafe.

# FRANCE TODAY

France is a republic with a strong national government. Its present government, called the Fifth Republic, has been in effect since 1958. The First Republic was established in 1792, during the French Revolution. Between 1792 and 1958, the structure of the French government changed a number of times.

The French Revolution was a historic turning point for the people of France. It introduced democratic ideals to France and ended the days of *absolute monarchy,* in which the king had almost unlimited power. The revolution paved the way for the development of a unified state, a strong central government, and a free society dominated by the middle class and the landowners.

Today, more than 200 years after the start of the revolution, the goals of the revolutionaries still ring out in France's national motto—*Liberté, Egalité, Fraternité* (Liberty, Equality, Fraternity)—and endure in its way of life. French society is no longer divided into the estates that gave the monarchy, clergy, and nobles all the political power. And the French people no longer suffer from widespread famine, as they did in the days before the revolution.

France now prospers as a *parliamentary democracy* with a strong, centralized government whose president is elected by the people. The president manages the nation's foreign affairs. The prime minister (also called the premier), who is appointed by the president, directs the day-to-day operations of the government. France's Parliament consists of two houses, the National Assembly and the Senate.

France is now a world leader in both industrial production and agricultural output. As a result, its people enjoy a high standard of living. Social security laws give French workers some protection from the economic effects of unemployment, illness, and old age. In addition, France is a member of the European Union (EU), a political and economic partnership of European nations. France is also a leader in the eurozone, the group of EU nations that use the euro as a common currency.

## FACTS

| | |
|---|---|
| Official name: | Republique Française (French Republic) |
| Capital: | Paris |
| Terrain: | Mostly flat plains or gently rolling hills in north and west; remainder is mountainous, especially Pyrenees in south, Alps in east |
| Area: | 212,935 mi² (551,500 km²) |
| Climate: | Generally cool winters and mild summers, but mild winters and hot summers along the Mediterranean |
| Main rivers: | Loire, Garonne, Rhône, Seine |
| Highest elevation: | Mont Blanc, 15,771 ft (4,807 m) |
| Lowest elevation: | Rhone River delta, 7 ft (2 m) below sea level |
| Form of government: | Republic |
| Head of state: | President |
| Head of government: | Prime minister |
| Administrative areas: | 22 metropolitan regions, 4 overseas regions |
| Legislature: | Parlement (Parliament) consisting of the Senat (Senate) with 343 members serving six-year terms and the Assemblée Nationale (National Assembly) with 577 members serving five-year terms |
| Court system: | Cour de Cassation (Supreme Court of Appeals), Conseil Constitutionnel (Constitutional Council), Conseil d'Etat (Council of State) |
| Armed forces: | 352,800 troops |
| National holiday: | Fête de la Federation - July 14 (1790) |
| Estimated 2010 population: | 62,558,000 |
| Population density: | 294 persons per mi² (113 per km²) |
| Population distribution: | 77% urban, 23% rural |
| Life expectancy in years: | Male, 78; female, 85 |
| Doctors per 1,000 people: | 3.4 |
| Birth rate per 1,000: | 13 |
| Death rate per 1,000: | 8 |
| Infant mortality: | 4 deaths per 1,000 live births |
| Age structure: | 0-14: 18%; 15-64: 65%; 65 and over: 17% |
| Internet users per 100 people: | 51 |
| Internet code: | .fr |
| Languages spoken: | French, rapidly declining regional dialects and languages (Provençal, Breton, Alsatian, Corsican, Catalan, Basque, Flemish) |
| Religions: | Roman Catholic 85%, Muslim 8%, Protestant 2%, Jewish 1%, other 4% |
| Currency: | Euro |
| Gross domestic product (GDP) in 2008: | $2.866 trillion U.S. |
| Real annual growth rate (2008): | 0.7% |
| GDP per capita (2008): | $46,807 U.S. |
| Goods exported: | Aircraft, automobiles, chemicals, machinery, wine |
| Goods imported: | Automobiles, chemicals, iron and steel, machinery, petroleum products, pharmaceuticals |
| Trading partners: | Belgium, China, Germany, Italy, Netherlands, Spain, United Kingdom, United States |

France is a major political and economic power in today's world. Its foreign policies affect millions of people in other countries. France also has a long and colorful history. It was one of the first European nations to overthrow its king and set up a republic.

France also has a well-developed educational system. French children between the ages of 6 and 16 must attend school. After five years of elementary school, students enter a four-year school known as a *collège*. After collège, students enter a *lycée*, a vocational or general high school.

General high schools prepare students for study at a university. Special schools called *Grandes Écoles* (Great Schools) equip students for high-ranking careers in the civil and military services, commerce, education, industry, and other fields.

Like other modern nations, France has its share of economic and social problems. Inadequate housing affects large numbers of immigrant workers and their families, who are crowded in city slums. Unemployment, especially among young people, is also a problem, while many older people on fixed incomes have financial troubles.

# ENVIRONMENT

From the snowy peaks of the French Alps to the rolling plains and sunny beaches of the Aquitainian Lowlands, France is noted for its splendid scenery. Each of the country's diverse geographical areas has its own natural beauty.

The Brittany-Normandy Hills of northwest France have low, rounded hills and rolling plains with some rocky land. This region is actually a peninsula that juts out into the Atlantic Ocean. Steep cliffs on the northern coastline along the English Channel plunge almost vertically into the sea.

The southern coast of the peninsula, which faces the Bay of Biscay, has both sandy beaches and rocky cliffs. Many charming fishing villages line both coasts. Farther inland, small dairy farms—as well as apple orchards and grasslands—dot the landscape, their fields protected from the sea breezes by thick hedges called *bocage*.

Northeast of the Brittany-Normandy Hills lie the fertile and productive Northern France Plains—a landscape of flat and gently rolling land, broken up by forest-covered hills and plateaus. This region includes the Paris Basin, or *Île-de-France*, a large, circular area surrounded by lush, forested hills and drained by the Seine and other major rivers.

The gentle, rolling landscape of the plains gives way to the forests of the Northeastern Plateaus. Here, France shares with Belgium the Ardennes Mountains and their large deposits of iron ore. This wooded region becomes a little more rugged to the southeast in the Vosges mountains.

The Rhine River, Europe's most important inland waterway, flows along France's eastern border through the rich farmland of the Rhine Valley. Trees and vines cover the higher slopes.

Farther south, between the Rhine and Rhône rivers, lie the Jura Mountains. Much of the valuable forestland that once blanketed the higher slopes has been cleared for pasture, and vineyards cover many of the lower slopes. The Juras form part of the boundary between Switzerland and France.

Red-roofed cottages nestle comfortably around the chapel of Sons Brancion in western France. The rustic charm of these old buildings amid the rolling fields of the lowlands typifies the unspoiled beauty of rural France. About 23 percent of the French people live in rural areas.

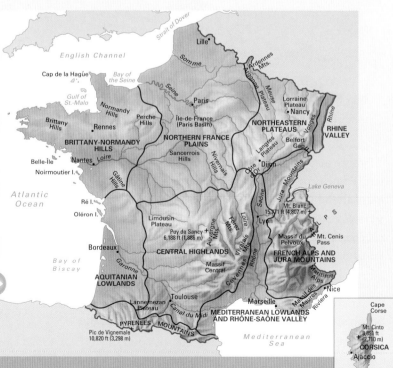

The varied landscapes of France include soaring mountains, deep river valleys, vast plateaus, and rolling plains. The Atlantic coast features steep, rugged cliffs, while the French Riviera on the Mediterranean shore provides miles of sun-drenched beaches. France's beautiful scenery attracts visitors from all over the world.

The French Alps begin south of the Juras, forming the border between France and Italy. Their majestic peaks provide some of France's most spectacular scenery.

Mont Blanc, which rises 15,771 feet (4,807 meters), is the highest mountain in the Alps. Thick woods and swift streams cover its lower slopes, but there is always a thick blanket of snow above 8,000 feet (2,400 meters). Mont Blanc also has huge valley glaciers, including the famous Mer de Glace (Sea of Ice).

West of the French Alps, the fertile land of the Mediterranean Lowlands and Rhône-Saône Valley produces fruits, vegetables, and wine grapes. The Central Highlands (Massif Central) of southeastern France features rolling hills and plateaus.

Along the country's southeastern border are the pine forests, rolling plains, and sand dunes of the Aquitainian Lowlands. The Pyrenees Mountains extend along France's border with Spain.

About 100 miles (160 kilometers) off the southeast coast of mainland France, the high, craggy coastline of the French island of Corsica rises out of the Mediterranean Sea. Sheep graze on the island's mountainous interior, and crops are grown in its narrow, fertile valleys.

The top half of Mont Blanc is almost always covered by a thick blanket of snow. Many fashionable ski resorts can be found on the lower Alpine slopes, which offer some of the world's finest skiing.

Workers harvest grapes near the village of St. Emilion in the Bordeaux region. The grapes are made into one of the region's many types of wine. France is the world's leading wine-producing country.

# PEOPLE

Though the people of France take great pride in being "French," the broad ethnic and cultural variety in France's population gives the country a unique and charming character. The influence of many traditions and cultures can be seen in regional differences around the country.

The provinces of France are home to many ethnic groups whose ancestors came to France from many different lands. Newcomers arrived during various periods of the country's history.

## Many languages and cultures

The people of Normandy trace their heritage to a group of Vikings from Scandinavia. These *Norsemen* raided and then settled along the French coasts and river valleys in the A.D. 800's and early 900's.

After colonizing this area, they invaded England under the leadership of William, Duke of Normandy, who felt he was the rightful heir to England's throne. Norman warriors crossed the English Channel and conquered the English forces at the Battle of Hastings in 1066. In time, the Norse influence spread throughout the British Isles.

The Norse heritage of the people of Normandy is still evident. They tend to be taller than most French citizens, with blue eyes and light-colored hair.

Along France's border with Belgium, many people in the province of Picardy speak the Flemish dialect of Dutch. Southeast of Picardy, the region of Alsace-Lorraine has a strong German heritage. This region has in the past been a prize in wars between France and Germany, and the people of Alsace-Lorraine have been part-German and part-French for hundreds of years. They speak a dialect that is closely related to Swiss-German.

High on the slopes of the Pyrenees Mountains live the Basques. Little is known about the ancestors of the Basques, except that they settled in the Pyrenees long before many other early settlers arrived in France. Today, their descendants speak a language called Euskera or Euskara. Many Basques

A group of Bordeaux residents enjoy a quiet game of cards under a shady tree. The Bordeaux region is noted for its vineyards, and the city is a leading center of wine shipping.

still wear the traditional red or black beret. The Basques are also famous for their lively, colorful folk dances.

Farther north along the Atlantic coast, some people in Brittany still speak Breton, a language related to Welsh. The area was settled by Celts, who came from what is now the United Kingdom in the A.D. 400's to 600's and named their new home Brittany (Little Britain). For many years, the Bretons tried to stay independent from France. They isolated themselves from the rest of France and developed their own culture.

Although life in Brittany today is more like that of the rest of France, many Bretons carry on the old customs and wear traditional costumes on special occasions. Breton women, for example, wear satin or velvet aprons and fine lace headdresses.

On the island of Corsica, most of the people speak a dialect similar to Italian. However, as in all these regions, French is taught in the schools.

## Immigrants to France

Today, many foreign residents add to France's ethnic mix. The largest foreign groups are people from Algeria, Cambodia, Italy, Laos, Morocco, Portugal, Spain, Tunisia, Turkey, Vietnam, and African lands south of the Sahara. Hundreds of thousands of refugees from former French colonies in Africa and Asia have arrived in recent years. Immigrants now make up about 7 percent of the country's population. The status of these immigrants is a controversial issue in the country. They may be the first to be laid off during slow economic times.

## Together as one nation

Although a number of dialects and languages are spoken throughout France, the official language is French. Now that French is taught in all the schools, fewer people in each new generation speak these local dialects or languages. The French government has now started to encourage its people to keep alive the age-old regional traditions.

Friends meet outside a bookstore in the Carré Rive Gauche (Left Bank) district in Paris. This old neighborhood is famous for its ancient universities, bookshops, and shops selling antiques and collectibles.

Folk dancers on stilts perform at a summer festival in Brive, a town in the province of Limousin. Such colorful festivals are held in many southern cities during July and August to entertain French families on vacation.

LIBRAIRIE ALAIN KOGAN

# EARLY HISTORY

Human beings have inhabited what is now France for more than 100,000 years. Little is known of the early inhabitants, but they left a remarkable reminder of their presence in the wallpaintings of the Lascaux Cave in southwestern France. These realistic paintings of hunters, bulls, horses, and reindeer date from about 15,000 B.C.

By about 500 B.C., tribes of Celts had settled in the region that is now France. The Romans, who began to invade the region about 200 B.C., called it *Gallia* (Gaul). Julius Caesar conquered the entire region between 58 and 51 B.C.

The people, who were called Gauls, soon adopted Roman ways of life. For example, the Gauls used the Latin language, which would later have an influence on the French language.

Many structures built during the Roman period still stand today. These include the large Roman bath at Cluny in Paris, as well as the Pont du Gard, an aqueduct near Nîmes, and amphitheaters at Nîmes and Arles.

## The Merovingians

In the A.D. 400's, the border defenses of the West Roman Empire began to weaken. Germanic tribes from the east—including Burgundians, Franks, and Visigoths—crossed the Rhine River and entered Gaul. After a series of battles with Rome and with the other Germanic tribes, Clovis, the king of the Salian Franks, founded the Merovingian *dynasty* (ruling family).

During the Merovingian dynasty, the rulers began to establish *manors* throughout France. These huge estates were governed by *landlords,* or *lords,* who offered military protection to the peasants, called *serfs.*

In the 700's, a system called *feudalism* began to develop. Under this system, a lord gave land to his noble subjects in return for military and other services. This land was called a *fief.* Some fiefs were quite large, such as the province of Normandy.

By the mid-600's, however, the Merovingian kings had become weak rulers. Pepin of Herstal, the chief royal adviser, gradually gained most of the royal power. His son, Charles Martel, increased the family's power and became king of the Franks in all but title.

Charles Martel's son, Pepin the Short, overthrew the last Merovingian ruler and became king of the Franks in 751. He founded the Carolingian dynasty, and his son, Charlemagne, became king in 768.

The cathedral at Rouen is a magnificent example of Gothic architecture. Many cathedrals were built in France during the Middle Ages, when architects believed that large, impressive buildings inspired greater faith.

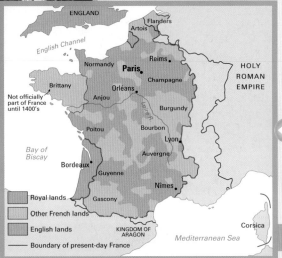

Royal power in France began to grow in the late 900's during the Capetian dynasty. Hugh Capet, the first of this long line of kings, firmly controlled only a small territory around Paris. Later Capetian kings enlarged the royal holdings, increased the power of the rulers, and established a strong, centralized government.

| | |
|---|---|
| c. 15,000 B.C. | Cro-Magnon people live in what is now southwestern France. |
| c. 600 B.C. | Greeks found the city of Marseille and name it Massalia. |
| c. 500 B.C. | Celts settle in France. |
| 121 B.C. | Rome gains control of southern France. |
| 58-51 B.C. | Julius Caesar conquers Gaul. |
| A.D. 400's | Germanic tribes (Franks, Burgundians, and Visigoths) enter Gaul. |
| 486 | Clovis, a king of the Franks, defeats the Roman governor of Gaul. |
| 507 | Clovis makes Paris the capital of the Frankish kingdom. |
| 751 | Pepin the Short becomes king of the Franks and founds the Carolingian dynasty. |
| 800 | Charlemagne becomes emperor of the Romans. |
| 800's–900's | Norse people conquer Normandy in northwestern France. |
| 814 | Charlemagne dies, and his kingdom is divided. Charles the Bald receives most of what is now France. |
| 987 | Hugh Capet is crowned king of France. |
| 1066 | William the Conqueror, a Norman duke, invades England. |
| 1100–1300 | French nobles fight in the Crusades. |
| 1100's–1400's | Cathedrals at Paris, Chartres, and Reims are built. |
| 1152 | Eleanor of Aquitaine marries Henry II of England, which begins English control of the duchy of Aquitaine in western France. |
| 1180–1223 | Philip II (Philip Augustus) rules France. |
| 1302 | Philip IV (Philip the Fair) calls together the first Estates-General, the ancestor of the French Parliament. |
| 1309–1377 | The popes live in Avignon. |
| 1337–1453 | France battles England during the Hundred Years' War. |
| 1346 | Battle of Crecy is won by the English in the greatest victory of the Hundred Years' War. |
| 1429 | Joan of Arc rescues Orléans from the English. Charles VII is crowned king at Reims. |
| 1453 | English are driven out of France. |

**Charlemagne (742-814)**

**Hugh Capet (940?-996)**

**Joan of Arc (1412?-1431)**

## Capetian kings

The first great Capetian king was Philip II, called Philip Augustus, who came to the throne in 1180. He developed Paris as a center of culture, government, and learning. Two of the city's greatest structures—the Cathedral of Notre Dame and the Louvre—were largely built during his reign.

The last Capetian king, Charles IV, died in 1328 without leaving a male heir to the throne. The ensuing dispute over the crown led to the Hundred Years' War (1337–1453) between England and France. At first, England won most of the battles. Later, the French king Charles VII regained power after the young peasant girl Joan of Arc led his troops to victory at Orléans. France eventually won the war.

## Charlemagne

Charlemagne, or Charles the Great, was the most famous ruler of the Middle Ages. Through a series of military campaigns, he expanded his territories far beyond what is now France. Charlemagne united these territories to create a great empire, and he revived political and cultural life in Europe.

After the death of Charlemagne in 814, most of what is now France went to his grandson Charles the Bald. By the late 900's, the feudal nobles gained more political power. In 987, they ended the Carolingian dynasty and chose Hugh Capet as their king. He started the Capetian dynasty.

The Pont du Gard is an aqueduct built by the Romans to carry water to the city of Nîmes in southern France. The ancient Romans laid out a network of roads and built many public buildings, including some that still stand today.

# LOIRE CHÂTEAUX

The Renaissance, a great cultural movement that began in Italy, reached France in the late 1400's. Francis I, who was king of France between 1515 and 1547, became very interested in Renaissance art and literature. He brought Leonardo da Vinci and many other Italian artists and scholars to France. Francis I also tried to surround himself with the work of the Italian Renaissance masters.

The most magnificent *châteaux* of the Loire Valley were built during this period. The châteaux were used as elegant hunting lodges and country residences by the French royalty and nobility.

Today, more than 1,000 of these châteaux dot the landscape along the Loire River, showing France's lasting contribution to Renaissance architecture. The architectural heritage of this peaceful, wooded region of the Loire Valley makes it one of France's most famous regions.

During the 1500's, Francis I's Château d'Amboise was a magnet for scholars, poets, musicians, and artists. In 1517, Leonardo da Vinci came to Amboise at the invitation of the king. Scale models of his most noted scientific inventions are in the collection at Le Clos-Lucé, the house where Leonardo spent the last two years of his life, just southeast of the Château d'Amboise.

Visitors often begin their tour of the Loire châteaux at the mouth of the Loire, in the Breton seaport of Nantes. A bridge over the river offers the best view of Château d'Ussé, perhaps the most romantic and fanciful of these Renaissance castles. It is said that Ussé inspired Charles Perrault, who once lived there, to write the fairy tale "Sleeping Beauty."

To the east of Ussé, Château d'Azay-le-Rideau, among the most beautiful and harmonious of the Loire châteaux, stands on an island in the Indre, a branch of the Loire. Its massive corner towers soar to the sky. Colorful presentations in period costumes take place at this château on summer evenings.

The largest of the Loire châteaux, Chambord, was built for Francis I beginning in 1519. Much of its furniture was destroyed during the French Revolution. Chambord became the property of the French government in 1932.

West of Tours is the Château de Villandry, famous for its terraced gardens. These Renaissance gardens have been restored according to the original plans dating from the 1500's. Visitors can admire three tiers of gardens, including an herb garden, as well as a lake and about 7,500 square feet (700 square meters) of ponds and basins, all restored to their former beauty. In the Garden of Love on the lowest tier, beds of flowers represent the many forms of love in beautiful designs.

East of Villandry lies the Château de Chenonceau. Smaller than most other castles of the time, Chenonceau has a light and graceful quality. Chenonceau was Henry II's gift to his mistress, Diane de Poitiers. But after Henry died, his widow, Catherine de Médicis, forced the woman to give up the château. Driven from Chenonceau, Diane de Poitiers took up residence at the Château de Chaumont. Today, Chaumont contains a fine collection of tapestries and Renaissance furniture.

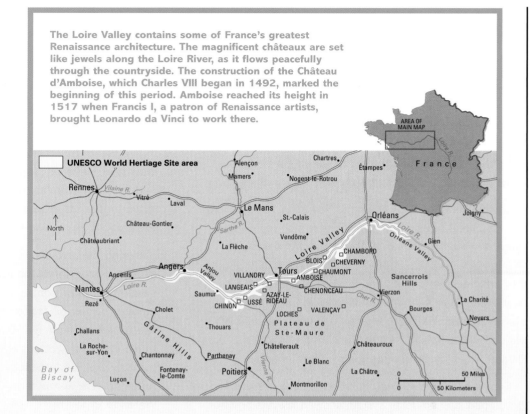

The Loire Valley contains some of France's greatest Renaissance architecture. The magnificent châteaux are set like jewels along the Loire River, as it flows peacefully through the countryside. The construction of the Château d'Amboise, which Charles VIII began in 1492, marked the beginning of this period. Amboise reached its height in 1517 when Francis I, a patron of Renaissance artists, brought Leonardo da Vinci to work there.

No visit to this land of fairy-tale castles would be complete without a journey to Chambord, the largest of the Loire châteaux. Chambord has 440 rooms, and its grounds were said to have once been as large as all of Paris. Here, in 1539, Francis I greeted the Holy Roman Emperor Charles V.

The builders of Chambord even altered the course of the Loire to enhance these 13,600 acres (5,500 hectares) of forests, gardens, lakes, and ponds. This wooded area, surrounded by a wall 20 mile (32 kilometers) long, is now a reserve for deer and wild boar.

During the Renaissance, the Loire châteaux were the center of social and intellectual life for the French nobility. However, an occasional dark deed occurred amid the splendor. In 1588, King Henry III's rival, Duke Henry of Guise, was murdered in the Château de Blois. It is said that the king himself, hidden behind a curtain, witnessed the killing. Today, Blois is most noted for the magnificent spiral staircase that many people believe was designed by Leonardo da Vinci.

Villandry is best known for its gardens. In the 1500's, rare species of vegetables, mostly imported from other countries, were grown in vegetable gardens near the château, where they could be guarded.

# HISTORY SINCE 1500

In the late 1500's, French Roman Catholics and Huguenots (French Protestants) fought a series of civil wars. In 1598, King Henry IV signed the Edict of Nantes, which granted some freedom of religion to the Huguenots. From the 1500's to 1700's, the power of the French kings and their ministers grew steadily.

Louis XIV came to the throne of France in 1643, when he was only 4 years old. He was king for 72 years—the longest reign in European history. Louis XIV believed in the complete political authority of the king. Although not well educated, he chose wise counselors, including Jean Baptiste Colbert, his minister of finance. Colbert promoted a strong economy. However, the cost of Louis XIV's wars of expansion, as well as the construction of his Palace of Versailles, drained the country's finances.

By the 1700's, a corrupt, bureaucratic government had developed, and an unfair system allowed lawyers and nobles to avoid paying taxes. Another series of expensive wars forced the government to impose even heavier taxes on the overburdened middle class. Out of this financial crisis, the French Revolution was born.

## The French Revolution

To gain support for more taxes, King Louis XVI called a meeting of the Estates-General in May 1789. The Estates-General was made up of representatives of the three classes of French society—the clergy, the nobility, and the commoners. The following month, members of the third estate—the commoners—declared themselves a National Assembly. They drafted a constitution that made France a limited monarchy, with a one-house legislature.

The French Revolution began on July 14, 1789, when crowds of Parisians stormed the Bastille fortress. It continued for many years as the revolutionary governments executed thousands of people they considered "enemies of the republic," including the French king, Louis XVI.

During the revolution, a young officer named Napoleon Bonaparte overthrew the revolutionary government and seized control of France. In 1804, he crowned himself emperor. Under his leadership, France developed a strong central government. By 1812, French forces had conquered most of western and central Europe. But the new territory was too large for Napoleon to govern. He was forced to give up the throne in 1814 and was defeated at Waterloo in 1815.

After 1815, the monarchy returned to France and tried to reinstate absolute rule. The attempt failed, and revolution erupted in 1830 and again in 1848. Napoleon's nephew, Louis Napoleon, crowned himself Emperor

The executioner shows the severed head of King Louis XVI to the mob in the Place de la Concorde in Paris on Jan. 12, 1793. Louis's wife, Marie Antoinette, was also executed, along with several thousand other people during the French Revolution.

Smiling Parisians greet the Allied troops that drove the Nazis out of Paris in 1944. German forces had occupied northern France beginning in 1940.

| | |
|---|---|
| 1572 | Thousands of Huguenots (French Protestants) are killed at the Massacre of Saint Bartholomew's Day. |
| 1589 | Henry IV becomes king and starts the Bourbon dynasty. |
| 1624 | Henry IV signs the Edict of Nantes, granting limited religious freedom to French Protestants. |
| 1643–1715 | Cardinal Richelieu becomes Louis XIII's chief minister and is considered the world's first prime minister. |
| 1789 | A crowd of Parisians storms the Bastille fortress, setting off the French Revolution and ending the absolute rule of kings. |
| 1792 | The First Republic is established. |
| 1799 | Napoleon seizes control of France. |
| 1804 | Napoleon founds the First Empire. |
| 1814 | Napoleon is exiled; Louis XVIII comes to power. |
| 1815 | Napoleon returns to power but is defeated at Waterloo. Louis XVIII regains the throne. |
| 1830 | Charles X is overthrown in the July revolution. |
| 1848 | Revolutionists establish the Second Republic. |
| 1852 | Louis Napoleon Bonaparte, nephew of Napoleon Bonaparte, declares himself emperor and founds the Second Empire. |
| 1870–1871 | Prussia defeats France in the Franco-Prussian War. The Third Republic begins. |
| 1870's–1914 | France establishes a vast colonial empire in Africa and Asia. |
| 1914–1918 | France fights on the Allied side in World War I. |
| 1939–1940 | France fights on the Allied side in World War II until defeated by Germany. |
| 1940–1942 | Germany occupies northern France. |
| 1942–1944 | The Germans occupy all of France. |
| 1944 | Germany is defeated by Allied forces. |
| 1946 | France adopts a new constitution, establishing the Fourth Republic. |
| 1946–1954 | France gives up French Indochina after a revolution in the colony. |
| 1954 | Revolution breaks out in the French territory of Algeria. |
| 1958 | A new constitution is adopted, marking the beginning of the Fifth Republic. Charles de Gaulle is elected president. |
| 1962 | France grants independence to Algeria. |
| 1969 | De Gaulle resigns as president. |
| 1981 | Socialist victories in presidential and parliamentary elections result in France's first leftist government since 1958. |
| 1993 | France becomes a founding member of the European Union (EU). |
| 1994 | A railway tunnel under the English Channel between France and England opens. |
| 1999 | France adopts the common EU currency, the euro. |
| 2011 | French President Nicolas Sarkozy plays a leading role in attempts to save the euro during a financial crisis. |

Cardinal Richelieu (1585-1642)

Napoleon I (1769-1821)

Charles de Gaulle (1890-1970)

Napoleon III in 1852. He was later overthrown by the people, and France was declared a republic once again.

## France after World War I

Heavy losses during World War I (1914–1918) weakened France. During the 1930's, a worldwide economic depression and the rise of fascist leader Adolf Hitler in Germany caused political unrest, leading to World War II (1939-1945). Hitler invaded France in May 1940. The northern two-thirds of France, including Paris, was soon occupied by German military forces.

On Aug. 25, 1944, the Allies defeated the Germans and liberated France. General Charles De Gaulle formed a provisional government and became president. Under his leadership, political and economic stability was restored. The French people voted to adopt a new constitution and in 1946, the Fourth Republic was formed. De Gaulle resigned over disagreements with the National Assembly, but he was called back to power as prime minister in 1958, when a new constitution established the Fifth Republic. Dissatisfaction with De Gaulle's policies led to student riots in 1968. He resigned in 1969.

In 1981, the French people elected their first leftist government since 1958. From the 1980's to the early 2000's, conservative and Socialist governments in turn guided France through the loss of its overseas empire and several economic crises. In 1993, France became a founding member of the European Union (EU), and in 1999, France replaced its currency, the franc, with the EU's common currency, the euro. In 2011, President Nicolas Sarkozy played a leading role in combating an economic crisis that threatened the euro. In 2012, voters displeased with Sarkozy's debt reduction program replaced him with Socialist François Hollande.

# PARIS

When Alexandre Gustave Eiffel designed his tower for the World's Fair of 1889, he could not have dreamed that it would become a symbol for one of the world's greatest cities—Paris, the capital and largest city of France. Today, the Eiffel Tower, rising 984 feet (300 meters), stands in the Champs de Mars, a park that was once a military training ground. From the top of the Eiffel Tower, visitors can enjoy spectacular views of Paris in all its beauty and charm, with its broad, treelined boulevards, historic buildings, and famous parks and gardens.

Paris is located in north-central France, in the heart of the Paris Basin. The Seine River flows through the city from east to west. The section north of the river is called the Right Bank. Busy offices, small factories, and fashionable shops are on the Right Bank. South of the river lies the Left Bank—the center of student life and a gathering place for artists.

Many historic monuments, churches, and palaces can be found throughout Paris. Each represents a different part of the fascinating, colorful story of this famous City of Light.

The Eiffel Tower is a symbol of Paris. The tower stands in a park called the Champ de Mars. The steel and iron structure was erected for a world's fair in 1889.

## Historic sights

At the western end of the Champs Élysées (Elysian Fields), Paris' best-known avenue, stands the Arc de Triomphe (Arch of Triumph). This huge stone arch was begun by Napoleon I in 1806 as a monument to his troops. The inner walls bear the names of 386 of his

Paris is a very ancient city, dating back to prehistoric times. Its modern street layout began taking shape in 1852 under Napoleon III. His designer, Baron Haussmann, leveled entire neighborhoods of narrow medieval blocks to create the network of wide, often diagonal avenues and boulevards.

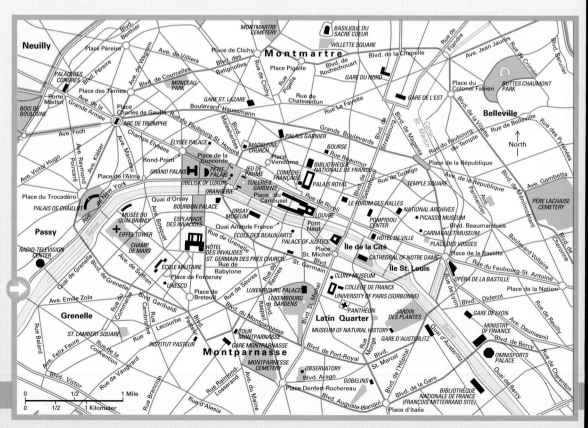

| | |
|---|---|
| 1572 | Thousands of Huguenots (French Protestants) are killed at the Massacre of Saint Bartholomew's Day. |
| 1589 | Henry IV becomes king and starts the Bourbon dynasty. |
| 1624 | Henry IV signs the Edict of Nantes, granting limited religious freedom to French Protestants. |
| 1643–1715 | Cardinal Richelieu becomes Louis XIII's chief minister and is considered the world's first prime minister. |
| 1789 | A crowd of Parisians storms the Bastille fortress, setting off the French Revolution and ending the absolute rule of kings. |
| 1792 | The First Republic is established. |
| 1799 | Napoleon seizes control of France. |
| 1804 | Napoleon founds the First Empire. |
| 1814 | Napoleon is exiled; Louis XVIII comes to power. |
| 1815 | Napoleon returns to power but is defeated at Waterloo. Louis XVIII regains the throne. |
| 1830 | Charles X is overthrown in the July revolution. |
| 1848 | Revolutionists establish the Second Republic. |
| 1852 | Louis Napoleon Bonaparte, nephew of Napoleon Bonaparte, declares himself emperor and founds the Second Empire. |
| 1870–1871 | Prussia defeats France in the Franco-Prussian War. The Third Republic begins. |
| 1870's–1914 | France establishes a vast colonial empire in Africa and Asia. |
| 1914–1918 | France fights on the Allied side in World War I. |
| 1939–1940 | France fights on the Allied side in World War II until defeated by Germany. |
| 1940–1942 | Germany occupies northern France. |
| 1942–1944 | The Germans occupy all of France. |
| 1944 | Germany is defeated by Allied forces. |
| 1946 | France adopts a new constitution, establishing the Fourth Republic. |
| 1946–1954 | France gives up French Indochina after a revolution in the colony. |
| 1954 | Revolution breaks out in the French territory of Algeria. |
| 1958 | A new constitution is adopted, marking the beginning of the Fifth Republic. Charles de Gaulle is elected president. |
| 1962 | France grants independence to Algeria. |
| 1969 | De Gaulle resigns as president. |
| 1981 | Socialist victories in presidential and parliamentary elections result in France's first leftist government since 1958. |
| 1993 | France becomes a founding member of the European Union (EU). |
| 1994 | A railway tunnel under the English Channel between France and England opens. |
| 1999 | France adopts the common EU currency, the euro. |
| 2011 | French President Nicolas Sarkozy plays a leading role in attempts to save the euro during a financial crisis. |

Cardinal Richelieu (1585-1642)

Napoleon I (1769-1821)

Charles de Gaulle (1890-1970)

Napoleon III in 1852. He was later overthrown by the people, and France was declared a republic once again.

## France after World War I

Heavy losses during World War I (1914–1918) weakened France. During the 1930's, a worldwide economic depression and the rise of fascist leader Adolf Hitler in Germany caused political unrest, leading to World War II (1939-1945). Hitler invaded France in May 1940. The northern two-thirds of France, including Paris, was soon occupied by German military forces.

On Aug. 25, 1944, the Allies defeated the Germans and liberated France. General Charles De Gaulle formed a provisional government and became president. Under his leadership, political and economic stability was restored. The French people voted to adopt a new constitution and in 1946, the Fourth Republic was formed. De Gaulle resigned over disagreements with the National Assembly, but he was called back to power as prime minister in 1958, when a new constitution established the Fifth Republic. Dissatisfaction with De Gaulle's policies led to student riots in 1968. He resigned in 1969.

In 1981, the French people elected their first leftist government since 1958. From the 1980's to the early 2000's, conservative and Socialist governments in turn guided France through the loss of its overseas empire and several economic crises. In 1993, France became a founding member of the European Union (EU), and in 1999, France replaced its currency, the franc, with the EU's common currency, the euro. In 2011, President Nicolas Sarkozy played a leading role in combating an economic crisis that threatened the euro. In 2012, voters displeased with Sarkozy's debt reduction program replaced him with Socialist François Hollande.

# PARIS

When Alexandre Gustave Eiffel designed his tower for the World's Fair of 1889, he could not have dreamed that it would become a symbol for one of the world's greatest cities—Paris, the capital and largest city of France. Today, the Eiffel Tower, rising 984 feet (300 meters), stands in the Champs de Mars, a park that was once a military training ground. From the top of the Eiffel Tower, visitors can enjoy spectacular views of Paris in all its beauty and charm, with its broad, treelined boulevards, historic buildings, and famous parks and gardens.

Paris is located in north-central France, in the heart of the Paris Basin. The Seine River flows through the city from east to west. The section north of the river is called the Right Bank. Busy offices, small factories, and fashionable shops are on the Right Bank. South of the river lies the Left Bank—the center of student life and a gathering place for artists.

Many historic monuments, churches, and palaces can be found throughout Paris. Each represents a different part of the fascinating, colorful story of this famous City of Light.

## Historic sights

At the western end of the Champs Élysées (Elysian Fields), Paris' best-known avenue, stands the Arc de Triomphe (Arch of Triumph). This huge stone arch was begun by Napoleon I in 1806 as a monument to his troops. The inner walls bear the names of 386 of his

Paris is a very ancient city, dating back to prehistoric times. Its modern street layout began taking shape in 1852 under Napoleon III. His designer, Baron Haussmann, leveled entire neighborhoods of narrow medieval blocks to create the network of wide, often diagonal avenues and boulevards.

The Eiffel Tower is a symbol of Paris. The tower stands in a park called the Champ de Mars. The steel and iron structure was erected for a world's fair in 1889.

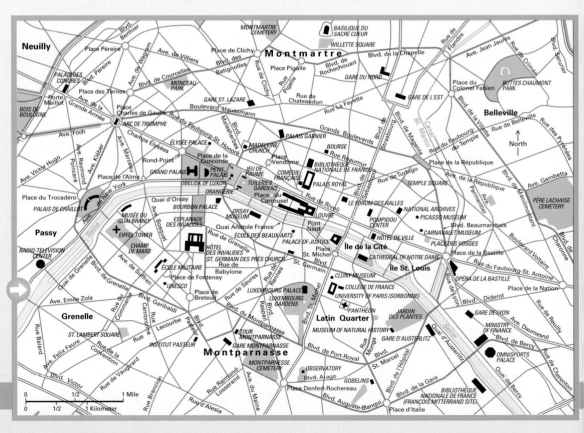

generals and 96 of his victories. After World War I (1914–1918), France's Unknown Soldier was buried beneath the arch.

At the eastern end of the Champs Élysées stands the Place de la Concorde (Square of Peace). It has eight statues, two fountains, and the Obelisk of Luxor, a stone pillar from the temple of Luxor in Egypt.

Built during the 1700's, the Place de la Concorde played a central part in a bloody chapter of French history. In this square, hundreds of people, including King Louis XVI and his wife, Marie Antoinette, were beheaded by the guillotine.

Across from the Place de la Concorde are the Tuileries Gardens—one of France's finest formal gardens and a favorite spot for Parisian children. They sail toy boats in the round lagoon fountains and enjoy Punch and Judy puppet shows in the park.

The Louvre, one of the largest and most famous art museums in the world, overlooks the Tuileries Gardens. Philip II originally built the Louvre as a fortress in about 1200. Today, it has about 8 miles (13 kilometers) of galleries and more than a million works of art.

The Louvre has especially fine collections of Egyptian, Greek, Asian, and Roman art. It also has an outstanding collection of paintings and sculptures of the 1800's, as well as decorative art. The museum's most famous works include the Greek sculptures *Venus de Milo* and *Winged Victory,* as well as Leonardo da Vinci's *Mona Lisa.*

The city's many historic churches include the magnificent Cathedral of Notre Dame. A fine example of Gothic architecture, the cathedral was one of the first buildings to use *flying buttresses* (arched exterior supports). The buttresses strengthened the walls and permitted the use of large stained-glass windows that let light into the interior.

## Economy

Paris is the chief financial, marketing, and distribution center of France. Many company headquarters and financial institutions operate in the city. Over half the nation's business is done in Paris. Jobs provided by national and local governments contribute greatly to the city's economy.

Paris is also France's transportation center. The city is served by three major airports, and it is the hub of a national railroad network.

Outdoor cafes and restaurants are a tradition in Paris. Local residents and visitors gather to enjoy a glass of wine or a meal during mild weather.

The Louvre is one of the world's great art museums. Much of the architecture dates from the 1500's and 1600's and contrasts vividly with the modern glass pyramid entrance.

# AGRICULTURE

The crops that grow on more than a third of France's land are an important part of the nation's economy. France, renowned for its wine, is a leading exporter of farm products.

## Crops of France

Wheat is the country's leading single crop. It is grown on large farms scattered throughout the northern region and Paris Basin.

Wine ranks as one of France's oldest and most famous agricultural products. About 2,000 years ago, the Romans cultivated vineyards in the southern region they called Provincia. Today, grapes are grown throughout France.

Grapes used to make high-quality wine come from several regions, including Alsace, Bordeaux, Burgundy, Champagne, and the Loire Valley. The Mediterranean region pro-

duces grapes used for cheaper wines, and grapes from southwestern France are used in brandy.

Other major crops grown by French farmers include beans, carrots, cauliflower, melons, mushrooms, onions, peaches, pears, peas, potatoes, sugar beets, sunflower seeds, and tomatoes.

## Raising livestock

About a fourth of France's land area consists of grassland used for grazing. The chief areas for raising livestock are the north and northwest, the Central Highlands (Massif Central), and the Alps. Brittany is a major center for the large-scale production of pigs and chickens.

Much of French farm income comes from meat and dairy products. Cattle, sheep, and lambs are the chief meat animals.

Fertile soil makes farming possible in almost every region of France. Regional specialties include cider and calvados brandy made from apples grown in Normandy and Brittany; poultry in Bresse; and lamb grazed on the salty meadows near the bay of Mont-Saint-Michel.

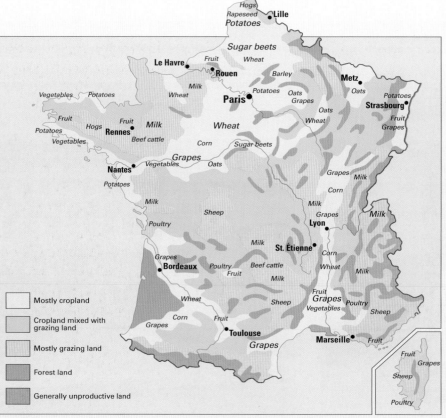

Rolling wheat fields in northern France produce much of France's grain harvest. Wheat is France's leading single crop. French farmers also grow barley, corn, oats, and rapeseed for livestock feed.

Gruyère de Comté cheese is one of three varieties of Gruyère, a popular cheese from France. Gruyère is made in the regions of Haute-Savoie and Franche-Comté. It requires very large amounts of milk and must cure for six months.

France produces butter and hundreds of kinds of cheese. Each region has its own specialties. Among the most famous of these cheeses are Gruyère from the French Alps, Camembert from Normandy, and Roquefort from south of the Massif Central.

## The new French farm

In the early 1900's, about half of France's workers labored on farms. Today, only a small percentage of the people work in farming. Between 1950 and 1980, about 500,000 French farms went out of business—mainly village farms that were too small to be profitable in modern markets.

Through the years, many of these small farms were abandoned. Others were bought by city dwellers who lovingly restored them for use as a second home.

The life of a French farmer today differs greatly from that of a village peasant in the past. Modern farms are much larger and more efficient. The people who operate them use good business sense as well as the latest techniques in production and marketing.

Lavender fields in Provence yield a sweet-smelling harvest. The lavender bush grows wild in the warm climate of the Mediterranean, but large fields are now cultivated. Lavender flowers and leaves are used to add fragrance to perfumes and soaps.

# INDUSTRY

France ranks as one of the world's leading industrial nations, with many of its workers employed in manufacturing. The nation's many automobile, chemical, and steel plants use modern equipment and production methods.

France is also involved in the manufacture of many other products, including sophisticated military and commercial airplanes, industrial machinery, and textiles. In addition, the country's fast-growing electronics industry produces computers, radios, televisions, and telephone equipment.

## The role of government

Most of France's industrial growth has occurred since the end of World War II (1939–1945). After the war, the French economy still depended chiefly on small farms and business firms. In 1945, however, the government introduced a program of industrial expansion and modernization.

High taxes and large government subsidies provided the necessary investment money. The French

France is Europe's leading producer of nuclear energy, which generates more than 78 percent of its electric power. France is also the world's largest exporter of electric power, primarily to Belgium, Germany, Italy, the Netherlands, and the United Kingdom.

France is one of the world's leading industrial nations. The country produces a wide range of products, including automobiles, chemicals, industrial machinery, perfume, steel, and textiles. Much of the country's success in industrial development is due to government planning. The state sets national economic goals and invests in research and development.

government also issued guidelines for private businesses. These reforms were made through a series of national plans, and they helped to make France an industrial leader in a relatively short time.

Most French businesses are privately owned. The government owns all or part of certain businesses, including some banks and steel companies. The degree to which the government controls businesses often depends on whether the Socialists or the Conservatives are in control of the French Parliament. Generally, when the Socialists have controlled the government, they have worked to increase government ownership of business. When Conservatives have been in control, they have sought to decrease government ownership.

## FRANCE'S HIGH-SPEED TRAIN

France's Train à Grande Vitesse (high-speed train), called the TGV, ranks among the world's fastest passenger trains. Developed by France's national railroad company in the 1970's, the first TGV began operating between Paris and Lyon in 1981. In 1989, an even faster TGV began running between Paris and cities in western France. The TGV runs mainly on straight tracks, which allows it to travel at speeds of up to 200 miles (320 kilometers) per hour. The TGV was designed to cope with increasing traffic on France's main railroad lines.

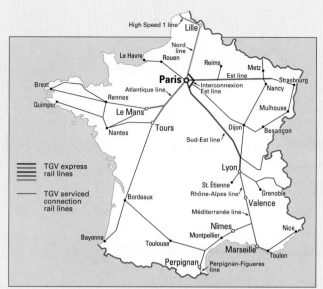

The TGV network spreads out from Paris, the hub of the national railroad system. Extensions to the TGV network serve Bern and Lausanne in Switzerland, as well as Nice, Toulon, and Marseille in France. The TGV Atlantique, opened in 1989, provides fast service to Rennes, Nantes, and Brest, as well as to Bordeaux in the southwest. The sleek TGV trains combine France's tradition of precision engineering with high technology.

## The Paris Basin

The Paris area is France's chief manufacturing, marketing, and distribution center. Industries in the Paris area include publishing and the manufacture of automobiles, chemicals, clothing, electronics, leather goods, and transportation equipment. Service industries—which include community, business, and personal services—are especially important to the Paris area.

## Industrial centers

Although Paris is France's major industrial center, there are factories in cities and towns throughout the country. For example, the silk weavers of Lyon produce some of France's most beautiful fabrics. The Lille-Tourcoing-Roubaix area also produces a wide range of textiles.

Major construction programs have brought industry to previously under-developed areas. Huge smelting plants have been built at Dunkerque, in northern France, and at Fos, near Marseille. Many plants serving the automobile industry have relocated to areas such as Dijon.

Industry in the Rhône-Alps region, with its main towns of Lyon, St.-Étienne, and Grenoble, has also greatly expanded. New businesses in the area include chemical, petrochemical, electrical, and electronic industries.

## European Union membership

France is a member of the European Union (EU), a political and economic partnership of European nations. The members of the EU have abolished all tariffs affecting trade among themselves. They have also set up a common tariff on goods imported from other countries.

In 1999, France and other EU countries adopted a common currency called the *euro* as part of an effort to create a more efficient, unified European economy. Euro notes and coins replaced the traditional currencies of these nations in 2002. France's currency had previously been the franc.

# THE ARTS

Throughout the ages, France has given the world some of its most brilliant artists, composers, and writers. That tradition continues today. French culture is a source of national pride and a symbol of the quality of life. Artistic expression is encouraged by the government through massive funding projects.

In recent years, the French government has sponsored the *Maisons de la Culture* (Cultural Centers). These multipurpose art centers, located in cities throughout France, are designed to bring culture to the provinces.

## Early art movements

During the Middle Ages, France's major contributions to the arts were the magnificent Gothic cathedrals built from about 1150 to 1300. The most important examples of these huge churches include the Cathedral of Notre Dame in Paris, as well as cathedrals in Amiens, Chartres, Reims, and Rouen.

Great triumphs in French architecture continued through the Renaissance, which reached its height in the 1400's and 1500's. The most impressive châteaux of the Loire Valley were built during this time. Literature also flourished during the French Renaissance. The writer Michel de Montaigne, for example, is considered by many to be the creator of the personal essay as a literary form in the late 1500's.

## Versailles and after

Baroque and rococo art developed in France during the 1600's and 1700's. Baroque art was large in scale and dramatic, while rococo art was smaller and more delicate.

France's greatest monument to Baroque art is the magnificent Palace of Versailles and its beautiful gardens. Versailles, the royal residence for 100 years, is now a national museum. The palace has about 1,300 rooms, and the grounds cover some 250 acres (101 hectares).

French Baroque music found its greatest expression in the operas of Jean Baptiste Lully and Jean Philippe Rameau. François Couperin composed music for the *harpsichord,* a stringed instrument similar to a piano.

The Palace of the Popes in Avignon was built in the 1300's, when Pope Clement V moved the pope's court from Rome to Avignon. Massive towers and walls that are 13 feet (4 meters) thick give the palace the look of a feudal castle.

The Hall of Mirrors in the Palace of Versailles exemplifies the drama and splendor of Baroque art. The graceful design of Versailles and its geometrically arranged gardens also reflect the classical themes of balance and order. Built by Louis XIV during the 1600's, Versailles lies about 11 miles (18 kilometers) southwest of Paris. Many of the rooms have been restored to look as they did when the palace was a royal residence. Visitors may also enjoy the many paintings and sculptures by famous European artists.

Classical art and drama also flourished during this period. Molière was the greatest writer of French comedy. Jean Racine and Pierre Corneille wrote tragedies.

The Age of Reason, also known as the Enlightenment, was a period of intellectual achievement in the late 1600's and 1700's. Writers such as Voltaire and Jean-Jacques Rousseau stressed reason and observation as the best way to learn truth. The Age of Romanticism, from the late 1700's to the middle 1800's, was a reaction against the Enlightenment. The romantics stressed emotions and the imagination.

The great French novelists during the Romantic period were Victor Hugo, Honoré de Balzac, Stendhal, and George Sand. The paintings of the Romantics

An open-air sculpture at the Pompidou Center intrigues a young visitor. The center provides a unique setting for exploring the vast collection of the National Museum of Modern Art.

The popular art movement known as Impressionism originated in France during the late 1800's. Edgar Degas was a leading French Impressionist artist, especially known for his paintings of ballet dancers.

were colorful and dramatic. Hector Berlioz was the greatest French Romantic composer. He gained fame for his large-scale orchestral works.

The middle and late 1800's brought a movement of realism and naturalism to France. Like many writers of the time, the novelist Gustave Flaubert tried to portray life accurately and objectively. Naturalism, an extreme form of realism, became popular in the vivid writing of novelist Émile Zola.

French painting soared to new heights in the late 1800's and early 1900's during the age of Impressionism. Impressionist painters tried to capture the immediate impression of an object or event. Leaders of this movement were Edouard Manet, Camille Pissarro, Edgar Degas, Claude Monet, and Pierre Auguste Renoir.

The contribution of French artists, writers, and philosophers has continued ever since. Outstanding literary figures during the early and mid-1900's included Paul Claudel, André Gide, Jean-Paul Sartre, and Albert Camus. Georges Braque and Pablo Picasso (who was born in Spain) helped shape modern art. Pierre Boulez and Olivier Messiaen were leaders in experimental music.

# CORSICA

About 100 miles (160 kilometers) off the southeast coast of France lies the island of Corsica—best known as the birthplace of Napoleon Bonaparte. Corsica has an area of 3,352 square miles (8,680 square kilometers). Its mountainous landscape, colorful villages, and ancient ruins have made this Mediterranean island a popular tourist attraction. Corsica is a French island and makes up two of the country's *departments* (administrative districts).

Although Corsica has been a part of France since the 1700's, the islanders do not feel a strong sense of French identity. Most Corsicans speak an Italian dialect rather than French, and the language differences further emphasize the separation between Corsica and mainland France.

Some Corsicans favor independence from France. Many others would prefer greater local control over the island's government. In 1982, the French Parliament created a Corsican regional assembly, elected by island voters. The regional assembly controls local spending, as well as the development of Corsica's economy, education, and culture.

## Early history

Tombs, *menhirs* (standing stones), and stone sculptures scattered throughout the island are all that remain of Corsica's prehistoric inhabitants. Beginning in about 560 B.C., the Phoenicians settled in Corsica, naming their new island home *Cyrnos.*

Since its early times, Corsica's location in the Mediterranean Sea has made it a place where many different peoples and cultures mixed. The island was conquered in turn by Etruscans,

Corsica is a mountainous island, reaching its highest point in Monte Cinto at 8,891 feet (2,710 meters). The western coast consists of rocky cliffs and narrow bays. Sandy beaches line all the coasts.

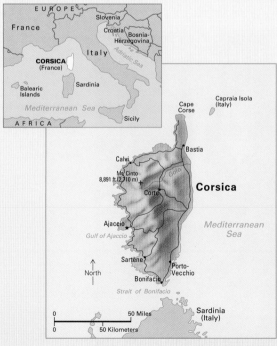

The mountain village of Corte, in the center of Corsica was the capital of the island from 1755 to 1769. A fortress built in the 1400's stands high above the town on a rocky ridge. Today, the fortress is occupied by the French Foreign Legion, one of the world's most colorful and gallant fighting forces.

Carthaginians, and Romans. At one time, it was controlled by Charlemagne, the most famous ruler of the Middle Ages.

About 1377, Corsica came under the control of the Italian city of Genoa. The Genoese sold the island to the French in 1768, who lost it to the British in 1794. Two years later, Napoleon sent an expedition to Corsica to reestablish French control. France has held the island ever since, except for a brief occupation by the British in 1814, and occupation by Italians and Germans during World War II (1939–1945).

## Land and economy

Corsica is one of the largest islands in the Mediterranean. Its coastline is high and rocky and has few natural harbors.

Corsica's rocky interior is covered with *macchia* (scrub). A variety of crops grow in its narrow, fertile valleys. Farmers raise olives, grapes and other fruits, grains, vegetables, and tobacco. The many olive trees, orange groves, vineyards, and flowers led Napoleon to once remark about his native island, "I would recognize Corsica with my eyes closed from its perfume alone."

Along the coast, people fish for sardines and hunt for coral. Since World War II, Corsica's main exports have been wool and cheese. Its fastest-growing source of income is tourism.

The French island of Corsica lies in the Mediterranean Sea between the southeast coast of France and the northwest coast of Italy. Ajaccio, the capital and largest city, is located on the island's western side.

# FRENCH GUIANA

A land of many rain forests and rivers, French Guiana lies on the northeastern coast of South America and ranks as the smallest territory on the continent. It is bordered by Suriname on the west, Brazil on the south and east, and the Atlantic Ocean on the north. Unlike its neighbors, French Guiana is not an independent nation, but rather an overseas *department* (administrative district) of France. It has a population of about 217,000.

Like other French departments, French Guiana has representatives in both houses of the French Parliament. French Guiana has two elected governing bodies, a general council and a regional council.

The interior of French Guiana has many natural resources, including rich soil, valuable timberland, and large deposits of *bauxite,* an ore used in making aluminum. However, these resources remain largely undeveloped. French Guiana relies heavily on financial assistance from the government of France to operate its government, support its industries, and pay for health care and other services. The chief industries include gold mining and the processing of agricultural and forest products. A shrimp industry is being developed. The leading farm products include bananas, cattle, corn, pineapples, rice, sugar cane, and yams.

## History

The French, who arrived in the 1600's, were the first Europeans to settle in what is now French Guiana. The territory became a French colony in 1667 and has been under French control ever since, except for a brief period in the early 1800's when it was governed by British and Portuguese military forces.

France began to send political prisoners to French Guiana in the 1790's, during the French Revolution. In 1854, Napoleon III established a formal prison system in the colony. Political prisoners were kept on Devils Island, an offshore isle, while other convicts were kept in prison camps in the towns of Kourou and Saint-Laurent du Maroni. About 70,000 people were held in French Guiana's penal colony between 1854 and 1945, when France closed the prisons, which had become notorious for their cruelty. In the 1960s, France turned the camp at Kourou into a space research center.

## People

A strong movement for independence from France developed in the 1980's. But most citizens of French Guiana wish the territory to remain a French department. Today, the French government is helping the department develop its economy and improve the life of its people.

French Guiana's three land regions consist of the coastal plain in the north, a hilly plateau in the center, and the Tumuc-Humac Mountains in the south. Rain forests cover much of the land, and more than 20 rivers flow northward to the Atlantic Ocean.

Buildings in the French country style line a street in Cayenne, the capital and largest city of French Guiana. The city was founded by the French in 1643.

About two-thirds of the people in French Guiana have full or partial African ancestry. Most are descendants of people who were brought to the territory during the 1600's and 1700's to work as slaves. Many are Haitians who moved to French Guiana in the 1980's.

Other groups include people of American Indian, Chinese, and European descent. The Indians were the first people to live in French Guiana. Today, they live in the rain forests of the interior. Most of the rest of the people live along the coast. A few thousand Maroons, who are descendants of Africans who escaped from slavery, also live in the rain forests and follow African tribal customs.

Most French Guianans speak French, the department's official language. Many Creoles, people with mixed black and European ancestry, also speak a dialect that is a mixture of French and English. Most of the people of the territory are Roman Catholics.

# FRENCH POLYNESIA

French Polynesia is a French overseas possession in the South Pacific Ocean. It is made up of about 120 islands scattered over an area about the size of Western Europe. The islands are divided into the Austral, Gambier, Marquesas, Society, and Tuamotu island groups. Papeete, on Tahiti—one of the Society Islands—is the territory's capital.

French Polynesia has a population of about 275,000 people. Most of the people are Polynesians. In 2004, France granted the islands the designation of *overseas country*. The islands' voters elect a territorial assembly, which in turn elects the president of French Polynesia. The islands also send representatives to the French Parliament. Tourism, agriculture, and fishing are important economic activities in French Polynesia. The chief products include *copra* (dried coconut meat), pearls, fish, and tropical fruits and vegetables.

An outrigger canoe floats over living coral and colorful fish in Bora-Bora's coral lagoon. Agriculture and fishing, mainly for tuna, provide most of French Polynesia's exports.

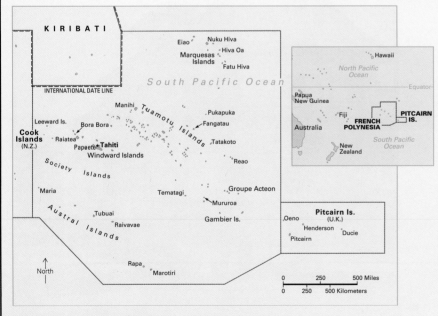

French Polynesia consists of about 120 islands scattered over an area approximately the size of Western Europe. Papeete, on the island of Tahiti, is the territory's capital.

## The Marquesas

About 10 rugged volcanic islands in the French territory make up the Marquesas, which lie about 900 miles (1,400 kilometers) northeast of Tahiti. The total area of the Marquesas is 492 square miles (1,274 square kilometers). The islands have steep mountains that drop sharply to the sea and fertile valleys with many streams and waterfalls.

About 8,700 people, mainly Polynesians, live on the Marquesas. Most provide their own food by farming or fishing. The chief crops are bananas, breadfruit, coconut, sweet potatoes, and taro. *Copra* (dried coconut meat) is the main export.

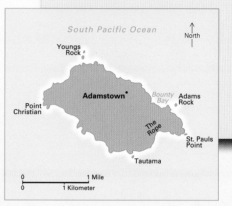

## PITCAIRN ISLAND

Pitcairn, the main island of a British dependency called the Pitcairn Islands group, was the home of the infamous mutineers from the British naval ship *Bounty*. In 1790, 9 British sailors from that ship settled on Pitcairn with 19 Polynesians—6 men, 12 women, and a young girl. About 50 people—mainly descendants of the *Bounty* mutineers and their Polynesian wives—live on the island.

The island, which covers only about 2 square miles (5 square kilometers), rises sharply from the sea. Its interior is rugged, but Pitcairn has fertile soil. Most of the people farm and fish for a living. Adamstown is the island's only settlement. The main crops are bananas, citrus fruits, coconuts, pumpkins, taro, watermelons, and yams. The people also sell hand-carved wooden figures to passengers of ships that stop at the island. The government gets much of its revenue by selling postage stamps that bear the words "Pitcairn Islands" to collectors.

Solitary Pitcairn Island lies just south of the Tropic of Capricorn and about 5,000 miles (8,000 kilometers) east of Australia.

**Stilt houses in Bora-Bora's lagoon are part of a tourist resort. Tourism is a major industry in the territory.**

## The Society Islands

This group of 14 islands lies southwest of the Marquesas. The island group's ancient volcanoes form many high peaks, making the land rough and mountainous. Some of the islands are low atolls used as fishing centers. The Society Islands cover an area of 613 square miles (1,587 square kilometers) and have a population of about 215,000.

Tahiti and Raiatea are the largest islands of the group, and Tahiti is the most populous. The capital of the Society Islands is the busy seaport of Papeete on Tahiti.

One of the most picturesque islands of this group is Bora-Bora. It lies about 170 miles (270 kilometers) northwest of Tahiti. A barrier reef with a number of low islets encircles it.

Most of the people are Polynesians or have mixed Polynesian and European ancestry. A Chinese population of several thousand controls much of the retail and shipping trade on Tahiti. There and on the other Society Islands, the tourist industry is the major source of income. Many rural people make their living by farming, fishing, and diving for pearls.

## The Tuamotu Islands

The Tuamotu group, made up of 76 reef islands and atolls, stretches across almost 1,000 miles (1,600 kilometers) of the Pacific Ocean. The islands themselves cover about 300 square miles (775 square kilometers), and have a population of about 16,000 Polynesian people. Destructive hurricanes sometimes strike the islands, which have a tropical climate. Like other low-elevation atolls, the islands often have shortages of fresh water. These shortages are due to low rainfall and a lack of lakes and rivers. Pearl and copra are the islanders' chief sources of income.

France tested nuclear bombs on Mururoa atoll from 1965 until the early 1990's. Scientific inspections in 1983 and 1988 revealed that the explosions caused environmental damage.